POLITICS AND METHOD

POLITICS AND METHOD

Contrasting studies in industrial geography

Edited by
Doreen Massey and
Richard Meegan

Methuen
London and New York

First published in 1985 by
Methuen & Co. Ltd
11 New Fetter Lane, London EC4P 4EE

Published in the USA by
Methuen & Co.
in association with Methuen, Inc.
733 Third Avenue, New York, NY 10017

Typeset by Folio Photosetting, Bristol
Printed in Great Britain
at the University Press, Cambridge

British Library Cataloguing in Publication Data

Politics and method: contrasting studies in
industrial geography.
1. Industry — Location
I. Massey, Doreen II. Meegan, Richard
338.09 HD58

ISBN 0-416-36250-8

Library of Congress Cataloging in Publication Data

Politics and method.
Bibliography: p.
Includes index.
1. Great Britain — Industries — Location — Addresses, essays, lectures. 2. Industry — Location — Addresses, essays, lectures. I. Massey, Doreen B. II. Meegan, Richard A.

HC260.D5P65 1985 338.6′042′0941 85-4845

ISBN 0-416-36250-8

Contents

Notes on contributors

Doreen Massey is Professor of Geography, The Open University, Milton Keynes

Richard Meegan is a Research Member, CES Ltd, London

Peter Lloyd is Director, Centre for Urban and Regional Industrial Development, University of Manchester

John Shutt is Senior Research Officer, Sheffield City Council (Employment Department)

Alan Townsend is a Senior Lecturer, Department of Geography, University of Durham

Francis Peck is a Senior Lecturer, Department of Geography and Environmental Studies, Newcastle-upon-Tyne Polytechnic

Stephen Fothergill is a Research Associate, Department of Land Economy, University of Cambridge

Graham Gudgin is a Senior Research Officer, Department of Applied Economics, University of Cambridge

Kevin Morgan is a Research Fellow, School of Social Sciences, University of Sussex

Andrew Sayer is a Lecturer in Geography, School of Social Sciences, University of Sussex

1

DOREEN MASSEY &
RICHARD MEEGAN

Introduction: the debate

For more than a decade now a number of debates have been taking place within industrial geography. The period has been one in which issues of the geography of industry – of the spatial form of industrial decline and growth – have often been at the forefront of wider political debate. And in the discipline of geography there have been long discussions – in conferences, seminar-rooms, books, journals and special reports – on regional policy and what form it should take, and on the rapid-fire series of policies for the inner-city in which 'solution' after 'solution' has been flung in the direction of the latest areas of industrial dereliction to gain political recognition. There have been debates about the relationship which should exist, if there is to be any chance of success in the declared aim of evening-out the unequal geographical, industrial and social development of the UK, between national economic policies and more specifically spatial ones; the relevance of spatial policies at all has been questioned. There have been arguments over the significance and behaviour of big corporations, variously called monopoly capital, multinationals and the meso-economic sector. And related to this has been the question of the degree to which, and the way in which, the changing internal geography of the UK is a product of the UK's own changing position within the international division of labour. New, or newish, problems of the quality of jobs, the geography of technology, of branch-plant economies and of external control, and new proposals for solutions, from science parks to enterprise zones; all have been on the agenda for debate within industrial geography.

At the same time, over the same period, and often in the same journals, we have been conducting a debate about theory and method. Having largely divested itself of the baggage of mathematical models and, at least in terms of their formal application, the theoretical models of (neo-)classical industrial location theory, British industrial geography went off down a number of paths in search of new approaches. Perhaps the longest established, and for long dominant approach, is that of extensive empirical analysis and the identification of location factors and common properties. This approach itself has changed in form over the years, possibly in part due to the debate with other approaches which were soon to arise to challenge it, and it is now based – as this volume demonstrates – on a more explicit and sophisticated exposition of method than has previously been the case. The other approaches which have arisen have themselves been varied. There has been a behavioural school shading off into purely empirical studies of individual corporations. And the period has also seen the development of a range of alternative approaches, structuralist, Marxist and realist, which have focused on setting changes in individual industries and companies within a wider framework which could take account both of underlying capitalist social relations and of the broader context of shifts in the national and international political economy.

These two debates, about the nature of the changes under way and about strategies towards them on the one hand, and about theory and method on the other, have sometimes merged and occasionally been explicitly related; but all too often they have been conducted in parallel. Time has too often seemed too short for self-conscious reflection on the nature of the relation between theoretical, methodological and policy perspectives. This omission has been problematical in a number of ways. It has led to misunderstandings of each other's positions (some of which we finally discovered in the seminar of which this book is the outcome!). It has led to situations where too much is, implicitly, being piled into one debate, and where a lack of distinction between the elements of difference has led to non-communication ('A' does not accept 'B''s formulation in the first place; 'B' finds 'A''s criteria for evaluating a good explanation utterly spurious, and so forth). It has also led to problems for students in trying to grapple with such a confusion of argument and to fathom out the, often only implicit, different levels of debate.

The issue of the relationship between theory, method, politics and policies is common to all the social sciences, and the debate which we present here has relevance beyond industrial geography – in economics, in sociology, in other branches of human geography. Ever since the 'social sciences' were recognized as such there has been fierce debate over methodology and the status of policy recommendations. In the early days there was perhaps much wider acceptance of the links between the two than now. The classical economists, for example, were not slow to draw out policies from their theorizing (Ricardo's decisive intervention in the debate over the Corn Laws being one notable example) and criticisms of their policies invariably involved a critique of their theories and methodologies. The subsequent rise of empiricism and the philosophy of logical positivism, however, brought in its train the so-called 'quantitative revolution' and the development of 'value-free' analytical techniques which had the aim of making a separation between theory, methodology and policy. In economics, this was epitomized by the writings of Lipsey and Samuelson (Lipsey 1971; Samuelson 1967), whose introductory textbooks still dominate the teaching of the subject. But this broad school of thought in the social sciences has come increasingly under attack, particularly from structuralist and realist critiques – an attack which has meant that the whole question of the link between methodology and politics is now very much back on the debating table.

This book is built around a seminar which was held in 1983. It was organized under the auspices of the then SSRC, now ESRC, as part of the programme of Doreen Massey's Fellowship in Industrial Location Research. The aim was explicitly to allow time for a small group of participants to discuss the range of issues around the question of the relationship between policies, politics, theory and method. The day did not produce agreement on the relations between these terms. Indeed it did not even produce consensus on the precise meaning of all the terms themselves. But it did allow us the space for real clarification of each other's arguments, and also, we found, for the relief of not being simply defensive. Most of us were self-critical about some past position or other, all of us reflected on the way in which our own approaches had developed, on the twists and turns they had taken and on the problems which had been faced, and sometimes overcome. There were points, certainly, in the debate where an impasse was reached, but as someone said at lunchtime, voicing the thoughts of us all, 'I wonder why we've never done this before?'

The central issues

The papers which were presented at the seminar approached the question of the changing geography of employment from a number of different methodological standpoints. We have picked out what we think are the most important features of these positions, and the main differences between them, in short prefaces to each chapter. The differences are certainly marked and wide-ranging. Major contrasts exist, for example, in the conceptualization of the underlying forces of change and the structural relations through which these operate. The different approaches also have contrasting expectations regarding the generalizability of research findings and very different views on the generation and 'testing' of research hypotheses. The extent to which the process of production is explicitly integrated into geographical analysis varies markedly between the different methods, and even where production considerations are directly addressed there remain widely divergent emphases and interpretations. And these differences in methodological approach also reveal themselves in the kind of information collected and in the specific research techniques adopted.

There are also marked differences in policy recommendations between the different approaches. They run the range from traditional incentives-based regional policy, indicative local economic planning and policies aimed specifically at influencing the employment and production strategies of major corporations, to strategies involving increased public ownership of production and control of investment and plans for non-profit production and alternative technology.

In this collection what we are specifically interested in is the extent to which these different policy recommendations are related to the different methodologies adopted. Indeed, is there any connection? Most of the contributors here accept that some interrelationship does exist, although this is not seen as being determinate or one-way, with methodology automatically deciding policy conclusions. Stephen Fothergill and Graham Gudgin, however, disagree with this view. While they accept that there is a clear link between methodology and research findings, they do not see any integral relation between methodology and policy recommendations. This view could be challenged on the grounds that if methodology and findings are linked and findings and policy recommendations in turn are connected, then logically there must be some connection, however slight, between methodology and research conclusions. Fothergill and Gudgin would reply, however, that what intrudes to rupture this simple connection is 'ideology'. In their view ideology has a key influence in determining both the issues to be studied and the recommendations to be drawn from research findings.

In the discussion, there was general agreement that ideological position has an important role to play in the use that is made of research findings. A recent classic example of this is provided by the reception given to the finding by David Birch that small firms were responsible for some two-thirds of net employment growth in the US (Birch 1979). As Fothergill and Gudgin note in their contribution to the collection (chapter 4) this work was eagerly seized upon in the US and in the UK by those ideologically committed to the promotion of small firms. Policies aimed at encouraging the growth of small firms were developed accordingly. Yet, as Fothergill and Gudgin also point out, a quite different interpretation of Birch's finding could have been made by those not sharing the same ideological predilection towards small business. Why was it not seen, for example, as an argument for developing policies to help large firms which dominate overall employment and which were clearly in difficulties?

More contentious in this context, however, is the argument that researchers of different political persuasions are drawn more to the study of some topics than others. Although we failed to resolve this issue in discussion, basic differences in interpretation were brought into the open. Thus, in one usage, and one on which we were all generally agreed, 'ideology' is taken to mean a set of ideas or beliefs promoting, but at the same time hiding, the interests of particular groups – like the small-business lobby just referred to. Yet lurking in the discussion was another view, on which there was much disagreement, namely that ideas and theorizing can be divided into two categories: the autonomous and 'value-free' and the 'ideological', and consequently false.

An important methodological issue which emerged was whether ideology (in either sense) influences where researchers attempt to break into the system of causation. One suggestion was that 'radicals' may be more likely to break in at the level of the system as a whole; less radical researchers, on the other hand, might accept the system as given and confine themselves to exploring causal relationships within it. While these different levels of analysis inevitably influence the nature of any policy recommendations in the sense that one (the radical) will come up with policies addressed more towards systemic level changes than the other, the suggestion was that this is because of ideology – through its influence on the selection of the level of analysis – rather than the research methodologies adopted once this choice has been taken.

In the discussion two dangers were identified in taking this argument too literally. First, it was argued by Doreen Massey and Richard Meegan that 'taking the system as given' does not remove the need to understand its specific nature. Even if research is quite prepared to take the capitalist system as given, it is still necessary to conceptualize it as *capitalist*. Breaking into the system of causation 'below the level of the system as a whole' does not change this – the parts of the system must still be conceptualized in terms of the specific nature of their social relations. Second, and what a number of us were agreed upon in the discussion, is the danger that different research methods might be seen as being more appropriate at different scales/levels of analysis. Thus Andrew Sayer and Kevin Morgan make a useful distinction in their contribution (chapter 6) between 'extensive' and 'intensive' research. The former, defined as relying on the use of aggregate statistics, surveys and statistical analyses, has been most commonly used in economic geography with the preoccupation with discovering general patterns of spatial change. Intensive research, in contrast, is increasingly being used to explore in detail how causal processes work out in specific cases. With its emphasis on abstraction rather than on the empirical generalization common to extensive research design, it is heavily dependent on non-standardized and qualitative analytical techniques. Sayer and Morgan and others, argued that it would be a mistake, however, to view these very different methods as being applicable at different levels of analysis: the extensive at macro-level, say, and the intensive at micro-level. Their differences lie in their explanatory frameworks, not in their analytical scope.

The question then arises as to whether the two research designs are complementary: does one (the extensive) set the agenda for the other (the intensive)? To be compatible they must share the same conceptual framework. The issue is most simply explored by concentrating on polar positions within extensive and intensive research. Extensive research design is often based on a taxonomic approach, aimed at identifying pervasive systematic trends in aggregate variables and exploring common features and relationships within these aggregate patterns. Having identified aggregate trends, the procedure then is to disaggregate these into separate components. Indeed, 'components of change' analysis has become a widely-used technique for this, with its disaggregation of employment change into such categories as plant closure, openings, *in situ* employment expansion and contraction, and locational transfer. A crucial question here is the explanatory status given to these components. There are some who would see them as in themselves explanatory. None of the participants in the seminar would take that position. It was argued, however, that they might be stepping-stones towards a subsequent explanatory stage; in other words that they might be the extensive analysis before the intensive.

There was disagreement as to whether this was, in most cases, likely to be successful. Taxonomic procedures such as these are essentially classifications by outcomes, rather than by causes. A 'closure', for example, is not a category with explanatory power in the context of the studies we were discussing. A closure is an outcome – a description of the cessation of production and employment at a particular site, but it does not explain *why* this occurred. As Alan Townsend and Franc Peck pointed out, closures are not everywhere the same, and can occur in very different economic circumstances. They can be the result of the failure of a company, or can be the outcome of the geographical transfer of production within a multi-plant company, as part of the latter's growth strategy. So what sense does it make in those two very different scenarios to say that job loss was *caused* by plant closure? Explanation, it was argued, lies in the underlying forces producing this outcome. Why did one company fail? Why did the other company have to transfer production geographically to maintain its growth? To the extent that extensive research remains with identifying and categorizing outcomes, then, there is a degree of compatibility between the two research designs, in the sense that extensive research may well uncover patterns which require explanation. Compatibility in terms of explanation, however, is a totally different matter. The categories which are produced out of extensive research may not be meaningful for intensive analysis.

There is also the question of conceptualizing the components, once roughly sketched out. There was disagreement over the importance of conceptualizing 'a new firm', for instance, as a social form with a particular structure of internal social relations. From the intensive corner Massey and Meegan argued that such conceptualization was essential to understanding causal processes. It is, however, admittedly a difficult task. Fothergill and Gudgin felt that it was anyway unnecessary – 'a trivial semantic problem'. Again, compatibility between intensive and extensive seemed elusive, certainly something which needs careful construction and cannot simply be assumed to exist.

An important factor in this compatibility is the way in which the different research designs conceptualize the relationship between factors. By isolating individual factors, extensive research tends to treat these different categories as separable phenomena which can be added together to explain the aggregate patterns being studied (a kind of additive causality). This procedure, to the proponents of intensive research designs, is fundamentally mistaken because factors in their approach are viewed as being structurally interconnected. Because of this interdependency, there is no simple way in which factors can be first disaggregated and then added together again. 'Structural' in this sense does not mean 'aggregate'. It means the form of explanation adopted, the fact that processes (not 'factors' or 'variables') are structured together (not added) to produce any one actual empirical outcome.

Moreover, because individual factors may be structured together their combination can, it is argued by Sayer and Morgan, and Massey and Meegan, radically alter the way in which each individually works. Thus extensive research would expect a single national cause (say, changing interest rates) to affect all regions equally, any differences being due to the addition of other factors which can be separately accounted for. Intensive research, in contrast, works on the assumption that the same national cause can produce very different effects in different regions/locations because of the way in which the factor in question is articulated in those locations in relation to other factors. Thus a single national policy, like regional policy for example, may well produce one effect in one situation and a completely different effect in another precisely because of the way in which different processes interact. In one set of circumstances changing product and process technology within an industry, and trade union militancy on existing sites, may well encourage plant closure and transfer of production elsewhere. Yet in another situation, the perceived militancy of organized labour might well discourage companies from transferring work elsewhere and result perhaps in new capacity being added to existing sites. Disaggregating the factors involved into, in this example, regional policy, technical change and trade union militancy would clearly not help in explaining the outcome. It is argued, from this position, that 'factors' need to be conceptualized as *processes* and structured together interactively rather than just added up. The qualitative relationships between parts of an explanation, in this view, are not amenable to explanation by statistically identifiable cross-effects between variables. This argument was disputed by Fothergill and Gudgin in the seminar, who argued that statistical methods were indeed available which could make allowances for such cross-effects, but that, anyway, conceptually this was not a problem. Both sides of the argument here stubbornly refused to concede, and in the end an agreement to differ on this point was the only way in which the overall discussion could proceed!

Another irresolvable difference between the two research designs became clear in the related discussion over the appropriateness of different methods at different spatial levels and the status of corporate studies. And the basic difference in the conceptualization of structural interrelationships already referred to lay at the heart of this disagreement. Proponents of extensive research designs would not deny the need for study of individual corporate behaviour, but in their view this is only necessary when the company concerned *is* the pattern to be studied – a company-town, for example. In such a situation there is no problem for any case-study of the behaviour of the corporation being idiosyncratic. But in other circumstances the idiosyncratic behaviour of individual firms needs to be ironed out to get at the pervasive trends, the general causes. In the extensive research design, these causes are empirically identifiable through common outcomes after idiosyncratic behaviour has been cancelled-out in the aggregate pattern. It is therefore necessary to study a large number of firms (the statistical law of large numbers) and consequently this research design is particularly appropriate at wider spatial levels (regional, national) where large numbers of firms are to be found. The reply, from the intensive research camp, is that this identification of common outcomes cannot be used as explanation because of the nature of structural inter-dependence. If the structural relationships between factors can alter the way in which each of them works, it is not possible to identify causality by looking for common outcomes. It is necessary to look at how those structural relationships operate. Real causes are to be found in necessary relationships and underlying forces. This was an argument made by a number of participants. Peter Lloyd and John Shutt stressed the need to look at the underlying mechanisms of capitalism though without expecting them to 'show up' in the behaviour of every individual corporation. Massey and Meegan, in chapter 5, identify inherent, necessary spatial implications of certain forms of production change, but recognize that these will be combined, in particular situations, with a range of contingent conditions to produce the real variability of actual outcomes. Sayer and Morgan, too, made a distinction between necessary relations, which are properties of objects, and the contingent conditions in which these relations operate. In other words, what the extensive research design views as idiosyncracy is in fact the complex combination of these necessary relations with contingent factors in the real world. While individual firms are clearly unique (they have, for example, their own products, organization of production, employment practices and industrial relations, marketing strategies and locations) they nevertheless fit into an overall structure of interdependencies – within a sector, within a national and international economy. For intensive research, what is necessary is research which can disentangle the necessary and contingent relations within this structure. The issue it has to face up to is the nature of generalizations that can be drawn from research findings on the necessary relations. Again the two different views were argued strongly, no agreement was reached, but a good time was had by all.

How to use this book

Our hope is that this book will be used and worked with as well as simply read as a sequence of papers. The value of the seminar was in the clash of ideas and the interaction between approaches as well as in the exposition of the approaches themselves. We have therefore tried to structure the book in such a way that these relationships between positions are highlighted and easy to follow.

The book does not present answers. As we have said, differences between the various positions presented began with the very definition of the terms and of the questions, and at a number of points, after all the discussion, we still agreed to differ. It would be naive, therefore, to expect there to be some meta-position which we could present and into which all the contributions could fit. Attempts to produce such a thing usually fail to recognize that they are themselves proceeding from one perspective rather than others. Necessarily then this book is a handbook. It is up to you to explore and judge the different positions, though we have structured the book to help you do this. The book is the basis of debate, not the magic key to the correct line.

The preceding section of this introductory chapter presented some of the main lines of debate, the central issues, around which much of the discussion revolved. In the chapters which follow we have done two things to bring out the differences and similarities on these issues between the individual papers. First, each paper has a brief résumé which concentrates on these issues and relates them to the positions taken in other contributions. Second, within the papers themselves major points, statements of position and turning points in the argument are *italicized* so, we hope, both clarifying the structure of the argument and making it easier to pick out, summarize and relate, the positions in different papers.

In the course of debating these central issues, however, a host of other questions were raised, cropping up at a number of times in the discussion and/or being raised by a number of authors in their papers. Some of these questions were the direct expressions of the central issues: the meaning of 'a case-study'; the validity of components of change analysis; the kind of data sources required (and considered valid) by different approaches; the question of whether interviews should be informally structured and interactive or structured and replicable; and so forth. Other questions led off from the central issues to open up other fields of debate: the meaning of 'external control'; the problems of defining localities and regions, for instance. In order to make it possible for readers to follow these sub-themes we have made an attempt to pick them out in a carefully structured index which focuses on issues rather than words and which, we hope, will aid in a deeper exploration of these sub-themes than might be possible in a single read.

We should, in other words, like the discussion papers to be used as a focus for debate and discussion, perhaps in seminars, just as they were in the original workshop. What is the difference between Lloyd and Shutt's argument and that of Townsend and Peck? Are the underlying relations identified by Lloyd and Shutt, Massey and Meegan, and Sayer and Morgan exactly the same kind of thing? What are the arguments for and against case studies – and how does your answer depend on your overall approach to explanation? Should interviews try and iron out differences in situations to enable replicability or make positive use of them to unearth more depth of explanation? How much point is there in the state intervening in big corporations – and what theoretical position does your answer imply? What attitudes should researchers take to conflicts of interest in the situations they study? Or, the one which kept rising to the surface throughout our discussions: what, anyway, is 'the regional problem' about which we were all so exercised? All of them raise the wider question which underlies it all: the role of the researcher in society.

2

Editorial introduction

Lloyd and Shutt open the debate by setting the empirical scene which has been both backcloth and laboratory for the discussions within industrial geography in recent years. Very graphically, they paint the picture of decline, of the loss of jobs both nationally and in the region which they take as their focus, the North-West of England. It is perhaps not without its ironies that this region which, as Lloyd and Shutt point out, in the mere four years from 1978 to 1982 lost over a quarter of a million jobs, was the birthplace of industrial capitalism.

The path the authors trace in their contribution is precisely to start from this description, eventually to trace what they see as the real causes of present problems in the underlying mechanisms of capitalism itself. The story they tell is one of their own development of their methodology. Their initial question was the one which faces us all: how to move from a broad description to an analysis of causes. Lloyd and Shutt argue that an investigation of employment aggregates does not in itself get the analysis close to explanation. But nor does, in itself, disaggregation of those aggregate statistics. They report, in this context, a disillusion with components of change analysis ('beyond establishing a more sophisticated accounting framework, the components of change approach can make little progress in informing the debate on causality'). Their proposed solution is an exploration of causality at two, linked levels. On the one hand, they explore the behaviour, and the causes of the behaviour, of the main agencies of change and of job loss in their region – the big firms, the 'prime movers'. This gives them a finer understanding of the mechanisms of change and of some of its proximate causes but, the authors argue, it is necessary to go behind the immediately motivating forces to identify 'some of the broad forces for change'. On the other hand, they explore the systemic-level changes which lie behind the behaviour of individual corporations. This exploration reveals a small number of major processes in which the corporations of the North-West region have been caught up. These include in particular the centralization and internationalization of capital, and the evolution of a new phase of the international division of labour, and the increasingly rapid pace, over the years, of technological change, leading to the problem of the transiency of any individual pattern of investment.

It is important to be clear about the nature of the link which Lloyd and Shutt establish between the individual corporations on the one hand and the more systemic-level processes on the other. The latter is far more than just a context for the former; it is part of the explanation. And yet it is not a deterministic or mechanistic explanation. In the paper the operation of the systemic forces is exemplified by references to corporations, but there could be no question of 'proving' their existence thereby, nor of running regressions to establish their importance. For the systemic-level causes do not appear as dominant, identifiable, 'factors' in each and every case; nor do they appear in the same form in each case – corporations vary in their nature and in their responses. This link between the two levels is crucial. As the authors themselves put it: in looking at the wider processes they were 'seeking to conceptualize rather than generalize . . . *both* to explore the variety of company behaviour *and* to relate it to broad structural concepts derived from an analysis of capital accumulation and the circulation process' (our emphasis – eds).

By consistently pushing themselves to live up to their criteria for explanation rather than description, the authors thus end up with a wide canvas, both in terms of the forms of social relations which lie behind the present devastation of the economy of the North-West of England, and in terms of the international spatial extent of the causal processes which must be addressed. Their policy conclusions, again, rigorously accept the implications of their own analyses. They are clear – as are most of the authors in the collection – that a return to 1960s-type regional policy is not on: 'the processes of change comprise far more than localized responses to adverse factor-cost comparisons at a time of shrinking markets'. It is necessary to step back. And this the authors do, beginning by questioning the very objectives of regional policy. If it is necessary to challenge market processes then so be it – the authors propose both increased levels of public expenditure and increasing social control over the investment process, in order both to increase the overall level of activity and to change its direction towards more socially important, less simply-profit-oriented, initiatives. Such policies can be pursued, both at national level, where spatial policy should be made integral to economic policy as a whole, and at more local level, where on the basis of their findings about labour markets, the authors propose the county rather than the region as the most appropriate scale for policy.

2

PETER LLOYD &
JOHN SHUTT

Recession and restructuring in the North-West region, 1975–82: the implications of recent events

Introduction: UK regional trends in the 1970s

This chapter examines the impact of the current recession on industry and employment in the North-West region. In what follows, we seek to analyse events within the manufacturing sector of a peripheral regional economy at a time of deep recession. It is, however, our primary intention to go behind the aggregate statistics in order to reveal something of the deeper processes of regional industrial change and their complexities. The analysis is, at this stage, far from complete and we set out here to offer a preliminary contribution to debate rather than definitive conclusions.

It it is possible briefly to summarize the broad national background against which UK regional and urban employment events in the 1970s have to be set, it can be couched in terms of rising worker participation rates (both young entrants and married women) set against low and sharply falling levels of net job generation as de-industrialization, recession and monetarist economic policies work their way through the nation. Between 1971 and 1978, for example, there was a net loss of more than three-quarters of a million manufacturing jobs in the UK economy with every major manufacturing order showing an overall net decline. Only the compensatory growth of jobs in services and a variety of government manpower support schemes prevented unemployment rising earlier and faster. Combined with and as a part-product of these structural shifts in the size and nature of the available employment pool, there were some dramatic shifts in the geography of UK employment.

As Keeble (1980) points out, while the period 1959–66 saw only ten of the country's sub-regions decline in manufacturing employment (for the most part the older conurbations), 1966–71 saw all the major industrial centres experiencing a reversal of their employment trends. Spatially, as Massey and Meegan (1978) show, the period after the mid-1960s saw a pattern of sectoral decline which was specifically regional in its impact. Employment growth, such as it was, became transferred to the periphery – the outer South-East, East Anglia, the East Midlands and parts of rural Wales and Scotland. Indeed, it has now become commonplace to draw attention to the urban-rural shift in manufacturing employment and to the role which this plays in underpinning unequal growth between regions. Fothergill and Gudgin (1982), for example, emphasize the incidence of employment decline in those regions containing major conurbations as compared with the relative growth experienced in those such as East Anglia, the East Midlands and the South-West, which contain no major urban agglomeration. The work of Champion, Gillespie and Owen (1982) supports this analysis of employment trends and shows how, on the onset of recession after 1979, the West Midlands and the North-West suffered the most serious relative losses. At the level of the conurbations themselves, there was, as Danson, Lever and Malcolm (1980) indicate, a rippling outwards of employment decline as it spread from the inner cities in the 1960s to the suburbs in the 1970s. Such limited employment growth as there was favoured service occupations and increasingly tended to benefit the free-standing towns on the urban margins.

North-West England: the acceleration of decline

Focusing more specifically upon the North-West, trends here during the middle 1970s were a microcosm of those occurring in the nation as a whole. The North-West, however, saw an acceleration of its trajectory of decline rather than a reversal of previous growth as was the case in the inner South-East and the West Midlands. Strictly in manufacturing terms, the North-West had the worst employment performance in the nation, with only the South-East losing more jobs in absolute terms between 1971 and 1978. In all, some 164,500 jobs in manufacturing were lost in net terms. The North-West also failed to share in the compensatory growth of services. The, nowadays mandatory, shift-share analysis reveals that a structural predisposition towards decline at the beginning of the 1970s became exaggerated by a poor performance component (Lloyd and Reeve 1982). The flood of inward investment fell to a trickle and variously active and passive phases of regional policy saw more intra-regional than inward industrial movement (Lloyd and Mason 1979). In the intra-regional context, the events and trends of the 1970s mirror those revealed by Fothergill and Gudgin (1982) for the nation as a whole – urban and inner urban decline together with limited but relatively positive growth both in services and manufacturing in the previously less industrialized parts of the region.

Faced with the emerging scale of job loss during the current recession, however, even the depressing employment performance of the region during the early and mid-1970s has to be seen as a relatively gentle period of downward transition. The acceleration of job loss since that time has been of a magnitude which sets it quite apart from previous trends. The full impact of the post-1979 recession in the region is yet to be measured but it is clear from preliminary work that *the industrial fabric of the region has been seriously undermined in ways which recovery from recession cannot necessarily be expected to repair.*

In the manufacturing sector alone, there has been a further net loss of the order of 235,000 jobs amounting to 24 per cent of the total 1978 manufacturing employment stock. While the traditionally declining textiles and clothing sectors have contracted even more sharply, they have been joined by the engineering, electrical and vehicle industries which, collectively, shed over one fifth of their job stock between 1978 and 1982. A substantial proportion of these losses came from industrial sectors whose expansion during the 1950s and 1960s formed the criterion for success in a regional policy dedicated to achieving diversification and a more favourable industrial structure. The service sector, which served in the past to offer some measure of compensation for manufacturing losses, also turned down. Far from offering a counter-balance to manufacturing decline, the services added to it losses of some 13,800 workers. Transport and the distributive trades accounted for the bulk of the observed service losses and counter-inflation policies, as public sector cash limits squeezed demand out of the regional economic system, led to a drop in public sector employment of 4.3 per cent. A slump in construction led the industry to lose more than 20 per cent of its job stock while the utilities saw only modest employment growth.

In total, over a quarter of a million jobs were lost to the North-West region during the period 1978 to 1982, a fall of 10.5 per cent on a base year which was itself the end point of a period of accelerating job degeneration. Some additional evidence for the depth of the recession in the region can be gained from available data on redundancies from the Manpower Service Commission's ES955 series. In particular, it is the scale of plant *closure* which can be revealed – the scrapping or relocation of industrial capital which an upturn cannot be expected quickly to replace. In all, some 2070 plants within the scope of ES955 declarations have closed in the period since 1975, with closure accounting for 46 per cent of all notified redundancies. If declarations by the same plants in their run-down prior to closure are accounted for, the real proportion of redundancies in retrenchments which culminated in closure is considerably higher.

The emergence of widespread mass redundancy has been explored recently by Townsend (1980a) and Martin (1982). Using shift-share analysis, both suggest that the high rates of manufacturing job loss in such major manufacturing regions as the North-West cannot be explained simply by factors related to industrial composition. This would seem to be confirmed by the across-the-board incidence of redundancy revealed by more localized study. Indeed, no recent shift-share analysis has found any support for the structural hypothesis. In the specific case of the North-West (and the West Midlands), Martin argues that the differential rate of job decline is a product of the fact that

> manufacturing tends to be characterized by larger-sized plants, higher labour costs relative to net output and a greater proportion of old and ageing capital stock than in the South and East of the country. (Martin 1982, 28)

It might also have been added that the residual 'performance' element of the shift-share analysis also contains within it hidden 'quasi-structural' factors such as the corporate and organizational characteristics of industry. As we shall go on to show, much of what happens in a region such as the North-West is powerfully associated not simply with the industrial sectors to which its capital stock is structurally assigned but with the 'structure' of trans-national and national companies' branch plants, divisional headquarters and corporate control centres into which North-West based operations are encapsulated. The incidence of redundancy within and across the region, while it has some clear sectoral elements and is also clearly some function of the age of capital stock,[1] is more clearly interpretable through a perspective which focuses on corporate responses to recession and restructuring at a time of emerging new process technology.

In summary terms, then, the North-West has suffered a continuous process of industrial restructuring during the 1970s and early 1980s. The economic base of the region has been fundamentally weakened and re-oriented in ways which the long-promised recovery will by no means redress. Entire industries have disappeared and both the traditional sectors and those of more recent vintage have closed plant and shed jobs at an alarming rate. For Merseyside and for those communities in Greater Manchester and North-East Lancashire dependent upon the textiles-clothing complex, factory and mill closures became an almost everyday occurrence during 1981 as their major sources of job opportunity rapidly disintegrated. Even the hitherto more buoyant new and free-standing towns of the region began to suffer major redundancies among their dominant employers. Against such a backdrop, the decision by the government drastically to cut back assisted area status to 'those areas of the country with the most intractable problems of unemployment' has been itself overtaken by a massive decentralization of unemployment and the widespread dispersion of the 'intractable' condition.

Changing perspectives on the causes of emerging trends

EVOLVING CONCEPTUAL FRAMEWORKS

One of the difficulties encountered in moving from a broad description of unfolding events to an analysis of causes and a subsequent prescription for policy is that the investigation of employment aggregates does not, of itself, provide a sound basis from which to proceed to an exploration of underlying processes. Aggregates and net spatial shifts are the product of a variety of lower order events – opening, closure, *in situ* expansion and contraction – which are themselves based upon the strategies of employers and new ways in which these impinge upon the use of labour in the productive process. As Massey and Meegan (1982) show, the labour market impact of corporate strategies for intensification and rationalization combined with the effects on workers of the application of new technology may give rise to a variety of complex outcomes both sectorally and spatially, rendering it impossible to 'read off' from employment shifts at the aggregate level the nature of the decision process from which they emanated.

Earlier work at the North West Industry Research Unit (NWIRU) (see Dicken and Lloyd 1978) had gone some way towards a recognition of the complexity of the underlying processes producing change by adopting a components of change approach to explore the disaggregated 'events' which lead to aggregate shifts. *However, it has become clear that, beyond establishing a more sophisticated accounting framework, the components of change approach can make little progress in informing the debate on causality*. Other work at NWIRU also sought to integrate a better 'accounting' of significant employment events with a more soundly-based perception of the nature of the corporate context. This was initially achieved by associating particular categories of events with pre-defined subsets of corporate enterprise, categorized by size and by sector. By associating events with ownership, it was, for example, possible to identify the importance of merger and acquisition as a potent force for change in the late 1960s and early 1970s. The assignment of ownership to regional manufacturing establishments also stimulated interest in the issue of external control and, at the other end of the scale, generated an exploration of the real attributes of the local indigenous enterprise (see Dicken and Lloyd 1978; Lloyd and Dicken 1979, 1982). What this enabled us to do was, however, more adequately to *describe* events, to *categorize* them. The still outstanding need to *explain* them has taken subsequent work in two directions. The one saw a retreat from such loosely conceptualized aggregates as 'externally controlled' or 'foreign-controlled' to a case-oriented examination of individual companies. This work began with the identification of the region's 54 manufacturing *prime movers* whose plants accounted for almost half of the region's shopfloor workforce, and continues with a close monitoring of their corporate behaviour up to the present. The other approach led to an explanation of the processes of capital accumulation and circulation, seeking to conceptualize rather than generalize in accordance with the recommendations of Sayer (1982a). In an attempt to cross the middle-ground between these two approaches, *we have sought to see the behaviour of the region's corporate prime-movers against a broad conceptual framework couched in terms of capital accumulation and the labour process, identifying the dominant enterprises (large capitals) with Holland's meso-economic sector and proceeding both to explore the variety of company behaviour and to relate it to broad structural concepts derived from an analysis of capital accumulation and the circulation process.*

An essential feature of the process we are attempting to describe, therefore, is that it is a dynamic one – an ongoing process of change in which the differential life chances of individuals and patterns of social relations within the region are seen to derive at any one time from the flow of job opportunity and the income associated with it. These are, in turn, related to changes in the stocks of households and capital goods and the contingent flows of population and capital which re-mould them over time. At any one time, under particular circumstances and in particular localities, gross investment may be at, above or below replacement levels and as the stock of capital goods changes so also does the level of full capacity employment and the occupational and social characteristics of job opportunity. With successive rounds of investment, capital replacement will rarely be by a good of identical character since new investment will embody technical change and reflect changing investment priorities both by private capital and the state. These shifts will serve to alter not only labour requirements directly but will also lead to a re-working of the flows of raw materials and intermediate goods contingent to the production process.

Under contemporary conditions, it is clear that these complex relationships are undergoing considerable change during a period when capital accumulation has been seriously disrupted. The ramifications of these changes are reverberating widely through the world economy – producing significant changes at international, national and regional levels. For the UK in particular, the period since 1979 has been characterized by a dramatic phase of capital restructuring and by fundamental change in the nature of those circulatory networks which both bind local and regional economies together and serve to integrate them with the national and international economy.

The precise outfall of the effects of restructuring is difficult to interpret and predict since, as Sayer points out:

> restructuring . . . does not happen in every sector simultaneously nor continuously in any one sector. The timing, form and place taken by restructuring cannot be known in advance precisely because it is affected by contingently-related conditions such as labour organization, political intervention and the development of technology. (Sayer 1982b)

What is, however, becoming increasingly clear is that the nature of emerging changes in the face of a globally integrated world economy so effectively articulated by the activities of multinational enterprise is making concepts such as 'locality' and the 'local' or 'regional' economy hard to handle. To some extent this reflects the weakness of locality as a theoretic object – a view recently expounded by Urry (1981) and Jensen-Butler (1982) among others. In essence, where locality is given, say in the concept of an administrative unit or travel-to-work area and where that pre-defined unit becomes the focus of study, this may result in the 'closing-off' of those higher order flows in the circulatory process whose configuration provides the motor force for local events and processes. The replacement of capital stock by those who exert control over this process is frequently conceived in world terms and the relative position of a nation or region is only interpretable against this global frame of reference. In the modern economy, it is the flow of *labour* which explicitly retains the local or regional dimension. Locality in this sense, though still a weak analytic tool, has more relevance, since the local labour market has both a restricted range, articulated in terms of journey-to-work flows, and a behavioural dimension embedded in evolved cultural notions of locality and community.

In any contemporary examination of the relationships between the level of economic activity, the labour market and the allocation of life chances, a clear tension, therefore, exists between the kinds of 'spaces' over which flows of money capital, intermediate goods and labour are to be evaluated. While the dynamics of the local labour market and its impact upon job opportunities can be investigated within the 'local' context, the causality of those events lies in the circulatory flows of the corporate sector, for which a considerably wider level of spatial resolution is demanded as the appropriate framework of analysis. Regardless of the problems of appropriate spatial definition, there seems, however, little doubt that the current process of restructuring is tending to increase rather than reduce inequalities between regions (see Dunford 1977; Massey and Meegan 1979, 1982; Dunford, Geddes and Perrons 1981).

For particular regions like the North-West, the subject of this paper, a complex and dynamic restructuring process of capital produces both general and particular outcomes. Among those to be anticipated from a general analysis of the prospective impact of the current phase of restructuring in a depressed peripheral region are a growth in the external control of its basic industry, the de-skilling of the traditional workforce and, in particular, as Dunford, Geddes and Perrons (1981) maintain, 'a reduction in the coherence and integration of its production complexes as regional plants become increasingly called upon to produce only parts of products or to perform part of the assembly process' – a process which generates few intra-regional industrial linkages. More specifically, such processes will reflect the region's *particular* sectoral configuration, its degree, type and strength of labour organization and the policy context through which central and local government have sought to manipulate its attractiveness to new rounds of investment.

REGIONAL EVENTS FROM THE PERSPECTIVE OF CORPORATE CASE STUDIES

It is in seeking to explore the links between corporate sector strategies and the nature of changes in local labour markets that the parallelism between evolving conceptual frameworks and empirical work on the region's prime mover companies begins to provide valuable insight into the complexity of the mechanisms driving regional events. *It is not that empirical work can be expected necessarily to establish the validity of the conceptual framework but that the findings of the one should enrich the understanding of the other.*

In a previous paper (Lloyd and Reeve 1982) we have identified those corporate actors whose decisions have been crucial to evolving trends in the re-orientation of the region's manufacturing capital stock during the period after 1975.[2] These major firms each employing more than 2500 *manual*[3] workers in the base year are listed in Table 2.1. In all, the 54 selected companies listed accounted for around 46 per cent of the region's manual workers and the 13 largest among them accounted for 25 per cent of the total. Indeed, if the full range of employment multipliers generated by the 54 companies through industrial linkage and regional income effects were to be estimated, their overall impact upon the concentration of employment opportunity would be considerably greater. Even allowing for the difficulties of arguing by extension that there is a high degree of industrial concentration in the North-West, it can be suggested that these same companies represent the chief agents for the articulation of new and replacement investment, intermediate goods flows and industrial labour demand. Their strategic investment decisions as well as day-to-day policies on sourcing and subcontracting have a profound effect on the nature of current and prospective future industrial activity in the region – particularly at a time when net inward investment by newcomer firms has fallen to a trickle. It is, however, in the employment sphere that our present concern lies – under circumstances where, at the onset of a period of major recession and restructuring, employment opportunity and access to wage-income for the region's residents rested in the control of a limited number of large multi-plant, multi-locational and, for the most part, trans-national companies.

More recent work now enables us to explore the way in which the prime movers of 1975 responded to rapidly changing conditions in the period of inflation and recession which followed. We can, for example, examine the pattern of aggregate employment change for the group 1975–80, while, at the same time, presenting more detailed analysis of the behaviour of individual companies as they struggled to confront an increasingly hostile economic climate.

Table 2.1 North-West regional prime movers 1975: control by headquarters location

Firm[1]	*Headquarters location*	*Rank*[2]	Main activities
Foreign multinationals			
General Motors Corporation	Minneapolis USA	4	Automobiles
Ford Motor Co.	Dearborn USA	10	Automobiles
Philips Gloeilampenfabrieken	Holland	13	Electrical
Heinz, H.J., and Co.	Pittsburgh USA	46	Food
UK multinationals/nationals			
A. *Controlled external to the region*			
Imperial Chemical Industries	London	1	Chemicals/textiles
Courtaulds Ltd	London	2	Textiles
GEC Ltd	London	3	Electrical engineering
Unilever Ltd	London	6	Food, detergents
British Insulated Callender Cables	London	7	Cables
Vickers Ltd[c]	London	8	Engineering
Plessey Co. Ltd	Ilford, Essex	9	Electronics
Hawker Siddeley Group Ltd[a]	London	12	Engineering
Dunlop Holdings Ltd	London	19	Rubber
Shell Transport and Trading Co.	London	21	Oil, petrochemicals
Lucas Industries Ltd	Birmingham	22	Vehicles and aircraft parts
Reed International Ltd	London	24	Paper, packaging
Stone Platt Industries Ltd	London	25	Textile engineering
Tube Investments Ltd	Birmingham	26	Engineering
British Aircraft Corporation[a]	Weybridge, Surrey	27	Aerospace
Metal Box Ltd	Reading	29	Packaging
Bowater Corporation Ltd	London	30	Paper
Wittington Investments (Ass. British Foods)	London	31	Food
Associated Biscuit Manufacturers	Reading	32	Food
Thorn Electrical Industries	London	33	Electrical
Imperial Group Ltd	London	36	Tobacco
Distillers Co. Ltd	Edinburgh	37	Whisky
Guest, Keen & Nettlefolds Ltd	Smethwick, West Midlands	39	Engineering
Rank Hovis McDougall	London	40	Food
BTR Ltd	London	41	Rubber
United Biscuits (Holdings) Ltd	Isleworth, Middx	42	Food
Smith & Nephew Associated Co.'s Ltd	London	43	Textiles, pharmaceuticals
Cadbury-Schweppes Ltd	London	45	Confectionary
Delta Metals Co. Ltd	London	48	Electrical
Chloride Group Ltd	London	49	Batteries
Bridon Ltd	Doncaster	50	Wire, fibres, plastics
Glaxo Ltd	London	53	Pharmaceuticals
Spirella Group Ltd[b]	Letchworth, Herts.	54	Textiles

Firm[1]	*Headquarters location*	*Rank*[2]	Main activities
B. *Locally controlled*			
Pilkington Bros.	St Helens, Merseyside	11	Glass
Turner & Newall Ltd	Manchester	14	Asbestos
Tootal Ltd	Manchester	16	Textiles
Ferranti Ltd[c]	Oldham	17	Electrical engineering
Ward & Goldstone Ltd	Salford	23	Cables, electrical
Cammell Laird Shipbuilders Ltd[d]	Birkenhead	28	Shipbuilding
Co-operative Wholesale Soc.	Manchester	36	Food
Renolds Ltd	Manchester	38	Power transmission
Rolls Royce Motor Holdings[c]	Crewe	44	Transport engineering
Scapa Group Ltd	Blackburn	47	Paper machinery
Vantona Ltd[b]	Bolton	51	Textiles
Mather & Platt[f]	Manchester	52	Engineering
C. *UK public sector*			
British Leyland	London	5	Trucks, cars
Ministry of Defence	London	15	Armaments
British Rail	London	18	Rail engineering
United Kingdom Atomic Energy Authority	London	20	Nuclear fuels
British Steel Corporation	London	34	Steel

Source: NWIRU Data Bank 1975.

Notes:

1 The NWIRU data bank classifies each manufacturing establishment by ownership using standard directories, e.g., *Who Owns Whom, UK Kompass.* A number of significant ownership changes have occurred since 1975 to this group, principally:

(a) the establishment of British Aerospace as a state-owned corporation in 1977 out of interests previously owned by Hawker Siddeley and British Aircraft Corporation ;
(b) the merger of Spirella and Vantona Ltd ;
(c) the acquisition of a 50% shareholding in Ferranti Ltd by the National Enterprise Board in 1975 and subsequent sale of this holding in May 1981 ;
(d) the nationalization of Cammell Laird and the formation of British Shipbuilders ;
(e) the merger of Vickers and Rolls Royce Motors in August 1980 ;
(f) the acquisition of Mather & Platt by Wormold International (Australia).

2 Ranking is in terms of descending order of shopfloor workers in 1975, as defined by NWIRU data, and includes all firms in the NW Economic Planning Region at that time employing over 2500 manual workers.

Table 2.2 Manual employment change in the 54 major companies, employing over 2500 manual workers in 1975

	1975	*1980*	*Manual employment percentage change 1975–80*
Foreign multinationals	40,631	29,990	− 26.2
UK multinationals/nationals (controlled external to the region)	263,864	196,764	− 25.4
UK multinationals/nationals (locally controlled)	66,688	58,831	− 11.8
UK public sector	42,389	35,172	− 17.0
Total 54 prime mover firms	413,572	320,757	− 22.4
Percentage share of total manual employment in prime mover firms	46.4	41.5	

Source: NWIRU Data Bank (provisional, subject to revision).

Note: At this stage, the NWIRU data bank underestimates the *in situ* change component and caution must be exercised in interpreting the data.

Perhaps the clearest aggregate characteristic of the group of key firms over a period which spanned recession, weak recovery and major slump was, not unexpectedly, their net decline in employment. In all, the 54 companies selected in the 1975 cross-section shed in excess of 100,000 manual jobs – a figure which the nature of the data source would render as an under-estimate of the real scale of loss.[4] Despite this, however, the key firms lost little of their dominance over employment opportunity in the region and industrial employment concentration in the original 54 companies remained largely unaltered. As Table 2.2 shows, the heaviest losses in the period 1975–1980 were suffered by the region's small group of foreign multinationals and by UK firms with non-local headquarters.

Both shed around a quarter of their job stock over the period, while public sector concerns lost 17 per cent. The indigenous group of firms with North-West headquarters fared less badly in employment terms but, as we shall go on to show, it would be premature to cite this as evidence either for their greater buoyancy in difficult times or for their potentially important future role as a source of job generation.

Nationally available data for the companies presented in Table 2.1 permit us to reveal something of the employment background against which the performance of North-West plants can be set. Space here prevents us from doing more than highlighting salient features, however. In broad terms, as would be anticipated, Table 2.3 shows a depressingly uniform pattern of corporate job loss. In most cases, this spans both the 1975–80 and the 1980–2 recession periods. Among the foreign multinationals with North-West plants, Ford remained remarkably stable in UK terms as well as within the region. Unfortunately, comparable General Motors data for 1982 was not available, but the overall indication is of 'moderate' job loss in both American motor manufacturers by comparison with the extremely large reductions at British Leyland. Among other foreign-owned companies, however, Philips, the Dutch multinational, has shed some 45 per cent of its UK employment over the period 1975–82 with almost all of this coming in the latest recession.

Among UK multinationals with industrial capacity in the North-West, the position of those with growing employment is easy to identify because of the relative rarity of the condition. For example, British Aerospace, Ferranti and the United Kingdom Atomic Energy Authority appear to have survived unscathed. In part, the first two have been able to ride out the recession longer and better than most because of the increase in defence expenditure and the substantial civil and military business placed before the downturn began in the mid-1970s. However, elsewhere in the domestic group of companies, the pattern of UK employment change is one of unremitting shrinkage. Two groups of companies can be identified: those experiencing some growth of employment in the 1975–80 period but which have now begun the job-shedding process. Ford, Shell, Chloride, Huntley & Palmer, BTR, ICL, Imperial Group, Thorn/EMI, Johnson and Firth Brown, Scapa Group, Transparent Paper, Bowater and Distillers all fall into this category, frequently as a

Table 2.3 UK employment change in the North-West region's major manufacturing firms, 1975–82

	UK employment		Percentage change	
Firm	*1975*	*1982*	*1975–80*	*1980–2*
Foreign multinationals				
General Motors Corporation	42,892	n.a.	− 2.5	n.a.
Ford Motor Co. Ltd	66,000	65,200	+ 15.2	− 14.2
Philips Gloeilampen Fabrieken	45,698	25,000	− 34.4	− 16.7
Heinz, H.J. and Co.	10,327[1]	8,527	− 2.2	− 15.6
UK multinationals/nationals, controlled external to the region				
Imperial Chemical Industries	132,000	74,700	− 36.1	− 11.4
GEC Ltd	171,000	145,000	− 10.5	− 5.2
Courtaulds Ltd	123,675	62,636	− 37.4	− 19.1
British Aerospace	35,767[2]	77,475	+ 112.2	+ 2.1
Vickers Ltd	30,690	18,500*	− 21.8	− 23.0
Unilever Ltd	91,394	73,252*	− 13.4	− 7.4
British Insulated Callender Cables	36,720	29,600	− 12.6	− 7.8
Plessey Co. Ltd	55,180	33,026	− 31.1	− 13.1
Shell Transport and Trading Co.	32,000	20,033*	+ 3.1	− 39.3
Lucas Industries Ltd	68,693	53,003	− 1.3	− 21.8
Metal Box Ltd	31,157	25,226	− 7.0	− 21.4
Chloride Group Ltd	10,097	7,823	+ 4.3	− 25.7
Stone Platt Industries Ltd	10,211	n.a.*	− 22.8	n.a.
Reed International Ltd	58,800	45,300	− 13.7	− 11.2
Wittington Investments (Associated British Foods)	77,663	72,419	− 7.3	+ 0.5
Huntley & Palmer Ltd (Associated Biscuits)	11,632	15,047*	+ 40.2	− 7.7
Hawker Siddeley Group Ltd	71,200	35,800	− 44.5	− 9.4
United Biscuits (Holdings) Ltd	27,000	30,000	+ 14.8	− 3.2
Dunlop Holdings Ltd	49,000	24,492	− 26.5	− 32.0
BTR Ltd	8,353	12,350	+ 81.1	− 18.3
International Computers Ltd	23,350	16,343	+ 9.9	− 36.3
Cadbury-Schweppes Ltd	28,475	23,384*	− 6.4	− 12.3
Imperial Group Ltd	87,428	82,700*	+ 2.0	− 7.3
Guest, Keen & Nettlefolds Ltd	74,048	40,000	− 23.4	− 29.5
Tube Investments Ltd	54,822	27,700	− 8.2	− 44.9
Thorn/EMI	82,404	78,083	+ 22.6	− 22.7
Rank Hovis McDougall	55,925	42,358	− 11.7	− 14.2
Smith & Nephew Associated Companies Ltd	11,508	7,729	− 25.4	− 10.0
Johnson & Firth Brown Ltd	12,517	8,623	+ 4.0	− 33.8
Glaxo Holdings Ltd	17,084	13,725	− 13.3	− 7.4
Burmah Oil	17,748	15,200	− 0.3	− 14.1

Firm	UK employment 1975	UK employment 1982	Percentage change 1975–80	Percentage change 1980–2
UK multinationals/nationals, locally controlled				
Turner & Newall Ltd	20,768	16,499	− 13.1	− 8.5
Pilkington Bros	23,000	18,280	− 1.7	− 19.1
Ferranti Ltd	16,651	17,850	− 0.6	+ 7.9
Tootal Ltd	19,813	10,995	− 11.3	− 37.6
Scapa Group Ltd	2,610	2,439	+ 25.1	− 25.3
Renold Ltd	11,334	6,177	− 26.0	− 26.3
Ward & Goldstone Ltd	6,415	4,059	− 9.7	− 29.9
Vantona Ltd	15,744	7,955*	− 42.1	− 12.7
Co-operative Wholesale Society Ltd	33,050	24,960	− 3.8	− 21.5
British Vita Ltd	3,277	2,792	− 8.7	− 7.0
Transparent Paper Ltd	1,279	1,232	+ 9.5	− 12.1
UK public sector				
British Leyland	164,354	80,600	− 17.9	− 40.3
Ministry of Defence (Royal Ordnance Factory)				
British Rail	251,627	170,397*	− 29.2	− 4.3
United Kingdom Atomic Energy Authority	13,025	14,350	+ 6.9	+ 3.1
British Shipbuilders	n.a.	66,320	n.a.	− 17.0
British Steel Corporation	223,000	103,700	− 18.8	− 42.7
Prime movers in 1975, which now employ less than 2500 manual workers				
Bowater Corporation	19,209	18,043*	+ 3.5	− 9.2
Wormold International (Mather & Platt)	n.a.	n.a.	n.a.	n.a.
Distillers Co. Ltd	20,100	18,125	+ 0.7	− 10.4
Delta Metals Ltd	25,000	14,775*	− 18.7	− 27.3
Bridon Ltd	8,232	4,029*	− 32.1	− 27.9

Source: Annual Reports, *Financial Times*, 9 December 1982.

Notes:

* In the absence of 1982 data figures quoted are for 1981.

1 This figure is for 1976.

2 This figure represents the average weekly number of employees in the British Aircraft Corporation Group in 1975, which was at that time part-owned by Vickers Group.

Individual company figures have to be interpreted with caution because they are affected by mergers and take-overs. Comparability is hindered because of changing definitions between companies, e.g. the treatment of part-time and seasonal employees which may or may not be included in the annual averages, and the fact that some companies report total employees at year end and others report averages at the time of the Annual Report.

product of acquisition and merger programmes. The second category contains the majority of firms in the region. These show a continuing pattern of job loss from 1975 onwards, with all but a handful experiencing a sharp acceleration in the more recent period. Most companies with key plant in the North-West demonstrate a devastating shake-out of jobs. Individual firms such as Shell, ICL, Dunlop, GKN, Tootal, Johnson and Firth Brown, and Ward and Goldstone have been forced to cut their workforces by at least one-third in the period since 1980 alone. These employment trends in the domestic group as a whole are, nevertheless, overshadowed by the massive scale of job loss in two key nationalized industries – British Leyland and the British Steel Corporation – both of which have reduced their workforces by over 40 per cent since 1980.

While *outcomes* with respect to employment in the region are relatively easy to describe, once again the processes at individual firm or sector level which gave rise to them are far more difficult to categorize. In part, the problem lies in the fact that many motivating forces are at work *simultaneously* in a general process of restructuring, thus defying simplistic attempts to establish a typology of dominant processes from extensive analysis of individual firm behaviour. In what follows *we attempt to identify some of the broad forces for change which appear to emerge in confronting the conceptual framework with the individual company case studies*. The outcome is no more than a first approximation and must be seen as a starting point rather than a basis for definitive conclusion.

The contemporary driving forces for industrial change within the region

THE DRIVE TO RATIONALIZE 1975–82

The specific weaknesses of the British economy are well known. Accumulation has been faltering since the late 1960s and falling profits have resulted in low investment levels (see, for example, Blackaby 1978; Glyn and Sutcliffe 1972; Singh 1977; CEPR 1981). The competitiveness of UK capital has been severely affected and the attempt to restore profitability has focused primarily on measures for reducing costs. Through the restructuring of production in various ways capital has struggled to maintain its competitiveness. Productivity gains have been sought by processes euphemistically described as 'slimming down' and the ratio of employment to output has been reduced in a drive to become 'leaner and fitter'. Massey and Meegan (1982) explored in some detail those strategies that capital adopts to achieve productivity gains. They identified three forms of production change: *intensification* of the production process (more output for less cost), *rationalization* to weed out inefficient plant and cut capacity and, where resources exist, *investment* to produce output gains from a lower cost base through technical change. Massey and Meegan examined how these processes evolved in the UK in a specific time period (1968–72) and from a sectoral perspective.

When examining industrial change from a North-West perspective over the period since 1975, it is sometimes tempting to argue that what is being witnessed is *wholesale rationalization* – indeed, the same could be said for the UK economy as a whole. Rationalization in Massey and Meegan's (1982) terms is simply conceived as the cutting of capacity with no new investment, frequently involving the closure of plant and the relocation of production to other parts of the UK. However, from a more realistic perspective based on corporate strategies in a globally integrated world economy, it becomes more difficult to adopt such discrete forms of categorization. Within a trans-national corporation, rationalization seen from a narrowly UK viewpoint may be taking place within the context of a worldwide reorganization of production. It may well be part of an integrated strategic plan involving new investment overseas within a corporate package which also contains some rationalization and the application of new technology at home and abroad. Rationalization in one sector of production may also be accompanied by diversification and new investment in another unrelated sector. In short, the choice of spatial and sectoral levels of resolution may give Massey and Meegan's categories a spurious discreteness which cannot be maintained in any more general context. At a time when recession is providing the driving force for capital to reorganize production and when the continuing centralization of capital takes place in the context of a communications revolution, corporate strategic planning and accumulation processes now operate at the world scale more than ever before. The connectivity of events at a variety of spatial scales both within and between sectors must not be overlooked.

Take, for example, the elements of Courtaulds' corporate strategy in the period since 1979. This may be summarized as follows.

(i) UK disinvestment in man-made fibres capacity resulting in the virtual elimination of man-made fibre production from the North-West.
(ii) Continuing rundown of plant and closure of mills in weaving and spinning in the UK as a whole.
(iii) Rundown of UK manufacturing capacity in the Consumer Products division, primarily in clothing manufacture.
(iv) Re-orientation of investment and production in the Fabrics division in order to increase the emphasis on value-added, well-designed products and drastically reduce dependence on commodity textiles and very basic lines.
(v) Re-configuration of the company base to reduce the textiles-clothing proportion by a programme of diversification and investment in paint and plastics manufacture, engineering and food products (broiler chickens).
(vi) A general internationalization of corporate operations entailing investment and acquisition.

Against such a complex background, Courtaulds has shed more than 61,000 UK workers since 1975, almost half of its original home workforce. To describe events in the UK simply as a product of corporate rationalization is clearly to miss the point. The Courtaulds case, while particularly revealing, is by no means unique. Strategic planning and corporate restructuring is taking place against a global backdrop and the bulk of those companies in the North-West whose recent activities have been subject to detailed analysis have been at pains to present themselves to shareholders as international in scope and diverse in nature. Internationalization, while a long-standing feature of UK capital, these days represents a favoured strategy in the production as well as the financial and commercial sphere. As such, it renders local (national) studies of local (national) events far too limited to provide a true window on underlying processes of industrial change.

THE CENTRALIZATION AND INTERNATIONALIZATION OF CAPITAL

It is clear that the dynamic of capital accumulation itself tends to produce an increasing concentration and centralization of production in capitalist economies. In essence, concentration is the process by which firms grow, continually seeking to plough back surplus profits, to cut costs, to obtain economies of scale and thereby continually to increase their market share. In the past, concentration of output in a small number of large firms in the UK has been the subject of investigation by Aaronovitch and Sawyer (1975), Hannah and Kay (1977) and Aaronovitch *et al.* (1981). On the basis of previous work by Hannah and Kay, Campbell (1981) reports that, if the rates of change experienced in the 1960s and early 1970s were to continue, the UK's one hundred largest firms would be producing 80 per cent of net manufacturing output by 1990. Already by 1976, just 87 companies in Britain produced over 50 per cent of UK exports, whilst 260 companies accounted for two-thirds (Panic and Joyce 1980). At the world scale, concentration ratios have been explored recently by Dunning and Pearce (1981) for a number of industry groups.

The concentration of capital is accompanied by the centralization of capital, that process by which firms amalgamate and grow. Mergers and acquisitions provide the main means by which centralization is achieved. At times of periodic crisis like the present, the weaker concerns fall out of production or face absorption by their stronger competitors. It is the tendency towards increasing concentration and centralization of capital which has spawned the rise of giant multinational firms and which leads them to articulate their activities systematically over global space. From the more localized perspective, the combined processes of concentration and centralization have ensured that regions like the North-West and their workforces have found themselves becoming *increasingly integrated into the global production networks of trans-national companies*.

Under these conditions, one of the most important driving forces for industrial change in previously industrialized peripheral regions is the emergence of that new phase in the internationalization of capital (described by Fröbel *et al.* 1980) in which companies have, in one way or another, been pressed into developing a new international division of labour. This has been achieved either directly through an active re-configuration of their own international operations or indirectly as domestic concerns have been forced to respond to the competitive pressures brought to bear by rival firms internationalizing their production facilities. *Key domestic and regional events must, therefore, be interpreted in the global context and against a background set by what Palloix (1975) describes as a 'new international phase of capital restructuring'.*

INTERNATIONALIZATION AND THE ISSUE OF LOCAL CONTROL

It is important to recognize that the emergence of global corporate planning is by no means the sole prerogative of foreign-controlled or recognized trans-national operators. Experience from corporate case studies in the North-West indicates that it is rapidly becoming a feature of those surviving locally controlled firms whose regional presence may sometimes be comfortably (if not necessarily accurately) associated with more localized forms of production integration.

In the textiles sector, for example, the recent activities of Tootal provide a direct challenge to the 'comfortable' interpretation of local control. One of the world's largest manufacturers of thread, textiles and clothing, Tootal was formed out of the 1960s merger movement involving English Sewing Cotton, Calico Printers and Tootal itself. At this time the company operated a decentralized divisional structure and, in the post-merger period, several organizational adjustments were made, but it was only with the arrival of the 1979 recession that a fundamental restructuring of the company's operations began. The effect of this has been to transform the company into a fully international concern, reducing eleven operations to four, each of which has a worldwide remit for the making and marketing of its products. As Leontiades (1974) points out, such volatile changes in organizational structure are becoming increasingly the norm for all those companies wishing to respond to a competitive and rapidly changing international configuration of production.

National and local outcomes which have followed this phase of corporate reorganization have been considerable. For example,

Tootal has shed 8818 jobs in the UK with 75 per cent of the loss appearing since March 1980. During the same period, overseas employment, particularly in associate companies, has grown by 8578. Within a strictly North-West context, Tootal shed 46 per cent of its 1975 labour force. More recently, Tootal has decided to transfer the headquarters of its thread division from Manchester to the United States, and there has been speculation that, having acquired Da Gama, a South African textile company, Tootal may in time utilize this as a centre of its global manufacturing operation (*Observer*, 19 October 1980).

While Tootal provides a prime example, similar instances of the thrust toward corporate internationalization and its impact through local rationalization and wholesale disinvestment are by no means rare. Renold, a long-standing Manchester-based chain and gear manufacturer, has been rationalizing its local and more traditional operations and the company's 1981 Annual Report is revealing for what it tells us about the pressures leading to internationalization and the national and regional multiplier effects set off by de-industrialization.

> Renold is now more an international and marketing group than a British company with overseas subsidiaries. In engineering products, particularly components, manufacturing production tends to follow the end-product market in which sales are made. This development is evidenced by the fact that Renold now makes more roller chain overseas than in the UK. Such a trend in other power transmission products seems inevitable unless the UK decline in the manufacturing of finished products is reversed. (Renold Annual Report, 1981)

Pilkington ranks as another locally controlled company engaged in a substantial international reconfiguration of its operations, particularly following its acquisition of Flachglas, a West Germany company, in 1980.[5]

INTERNATIONAL RECONFIGURATION AND THE REGION'S UK-HEADQUARTERED COMPANIES

Together with these changes overtaking locally controlled major enterprises, there has been a noticeable increase in the pace of international reconfiguration among UK-domestic companies in the North-West. Such companies as Dunlop, Thorn, Lucas and BTR can be recognized as dismantling particular North-West activities, while, at the same time, investing in related activities abroad. As early as 1980, for example, BTR, the rubber and plastics technology group, was reporting that 'from being a predominantly UK company in the early 1970s, BTR has built up both exports and overseas subsidiaries, so that over 60% of profits are now derived from overseas sales' (March 1980).

Dunlop, as Lane (1982) shows, also provides an interesting case. While the company was engaged in closing its Speke tyre plant, a £4 million investment programme was announced by the company in its Malaysian operations. Simultaneously, the activities of its Burnley golf-ball plant were transferred to the United States. Lane suggests that the company operates a policy of diversifying into new activities and using its UK base to pilot-test production prior to licensing manufacturing overseas, primarily to areas where labour is cheaper. In this respect, his judgement on the optimism which greeted the investment of £3 million in an injection moulding system for the production of wateproof boots in Liverpool is that 'this technology, although capital intensive to that being replaced, remains labour intensive. This suggests that once it has been "proved" it too will be exported. Remarkably similar tendencies obtain in the sports goods division' (1982, 10–11).

The tendency towards the creation of a new international division of labour through the current phase of the internationalization of capital by no means implies a simple transfer of production to areas where cheaper labour power can be obtained. Companies are seeking to increase their competitiveness in established markets in industrialized countries as well as opening up new markets. Thus a number of the North-West's prime mover companies have been engaged in strategic acquisition programmes, in particular in the USA.

Plessey Telecommunications, which has a strong regional presence, has adopted, since 1975, a strategy of rationalizing its telecommunications activities in response to the development of new telephone exchange technology. Simultaneously, the company was disinvesting in loss making consumer electronics and channelling investment towards the acquisition of high technology companies. While developing the manufacture of System X in the North-West, the company acquired Stromberg Carlson from General Dynamics in order to facilitate its sales of digital switching equipment to independent telephone companies in the USA. This acquisition will give Plessey the capacity to manufacture System X in the USA if necessary. Something of the logic behind this is explored in a recent *Financial Times* report:

> Plessey already has excess capacity in its British plants so the idea of starting System X production in the US may seem paradoxical. But telecommunications equipment customers all over the world are increasingly insisting that their suppliers manufacture locally and provide on the spot service and support. (*Financial Times*, 10 June 1982)

As with Plessey, firms such as GEC and ICL have increasingly found it necessary to buy a North American manufacturing base in order to obtain access to new technology and a launch pad from which to internationalize their operations. ICL, for example, acquired the Singer Business Machines Corporation and the Cogar Corporation USA in 1976 in order to expand its overseas marketing and acquire a USA manufacturing base (ICL Annual Report, 1976). In addition, the acquisition of SBM gave ICL access to Italy, Spain, Norway and Finland where previously no operation existed. The acquisition helped ICL in its move from electro-mechanical to electronic devices and in the transition from discrete to integrated circuitry, and led to the transfer of the System 10 product line from the USA to Letchworth. Subsequently, however, with the company's financial crisis in 1981, large scale rationalization saw the closure of ICL's Winsford and Plymouth Grove plants in the North-West. This has been accompanied by a survival strategy which focuses on buying in new products, thus avoiding high research and development costs. As Duncan (1982) points out, the joint venture with Fujitsu, one of the largest Japanese computer manufacturers, has given ICL access to more advanced chip technology. In parallel with this, the acquisition of new electronic telephone exchanges from Mitel Corporation of Canada and a deal with The Three Rivers Corporation USA will allow ICL to market telephone equipment, and manufacture and sell the PERQ personal computer workstation. In the case of ICL, then, internationalization is being undertaken by means of collaborative joint ventures developed as a strategic response to the company's

Table 2.4 Index of change in UK overseas employment levels in the North-West region's major manufacturing companies, 1975–82 (1975 = 100)

Firm	*Index of change* Domestic	Overseas	*UK absolute employment change*
(a) *Decline in both UK and overseas employment*			
Ford Motor Company	99	97	– 800
Courtaulds Ltd	51	87	– 61,039
Imperial Chemical Industries	57	92	– 57,300
Vickers Ltd	60	88	– 12,190
Unilever Ltd	80	95	– 18,142
Plessey Co. Ltd	60	87	– 22,154
Metal Box Ltd	81	87	– 5,931
Reed International Ltd	77	40	– 13,500
Cadbury-Schweppes Ltd	82	73	– 5,191
Glaxo Ltd	80	99	– 3,359
Burmah Oil Ltd	86	65	– 2,548
Renold Ltd	54	72	– 5,157
Bowater Corporation Ltd	94	68	– 1,166
Delta Metals Co. Ltd	59	42	– 10,225
(b) *Decline in UK employment and growth in overseas employment*			
GEC Ltd	85	126	– 26,000
British Insulated Callender Cables	81	130	– 7,120
Shell Transport & Trading Co.	63	113	– 11,967
Lucas Industries Ltd	77	165	– 15,690
Dunlop Ltd	50	101	– 24,508
International Computers Ltd	70	153	– 7,007
Guest, Keen & Nettlefolds Ltd	54	110	– 34,048
Rank, Hovis, McDougall	76	101	– 13,567
Smith & Nephew Associated Companies Ltd	67	114	– 3,779
Turner & Newall Ltd	79	112	– 4,269
Ward & Goldstone Ltd	63	129	– 2,356
Chloride Group Ltd	69	111	– 2,274
Pilkington Bros	79	129	– 4,720
Tootal Ltd	55	128	– 8,818
Scapa Group Ltd	93	148	– 171
British Leyland Ltd	49	134	– 83,754
Imperial Group Ltd	95	463	– 4,728

Firm	Index of change		UK absolute employment change
	Domestic	Overseas	
(c) *Growth in both UK and overseas employment*			
United Biscuits Ltd	111	111	+ 3,000
BTR Ltd	149	205	+ 3,997
(d) *Growth in UK employment but decline in overseas employment*			
Huntley & Palmer Ltd	129	42	+ 3,415
Ferranti Ltd	107	22	+ 1,199

Source: NWIRU records, company reports, direct contact with companies.

financial crisis. While there is little direct overseas investment involved, there has been a significant shift of emphasis with the company's UK operations becoming more oriented towards sales, marketing and software enhancement rather than manufacture of the products themselves. While the strategy may promote corporate survival, its regional impact will be substantial as will its effects on some grades of labour at present employed by the company.

In Table 2.4 we present what evidence is so far publicly available on the overseas employment shifts of the top 54 companies operating in the region for the 1975–82 period. Whilst the data must be treated with caution, they do provide one indication of the strength of overseas activity (particularly since 1980). In particular, Table 2.4 provides an index which compares UK employment shifts with overseas employment shifts. While it is not possible to read off from these data the true extent of the internationalization of production in this group of companies, the index does provide an indication of those firms pursuing vigorous overseas acquisition, investment and development programmes.

INTERNATIONAL RE-CONFIGURATION BY THE FOREIGN TRANS-NATIONAL AND ITS REGIONAL IMPACT

The latest stage in the internationalization of UK capital which we have been able only briefly to explore is taking place under circumstances where leading foreign trans-national companies are also engaged in a renewed drive to develop integrated global production systems. General Motors and Ford, both with key plants in the North-West, are, for example, currently locked into a corporate strategy to develop the 'world car' and 'world truck' as a means of achieving greater economies of scale and reducing R & D costs and marketing duplication. In the case of General Motors 'Interchangeable components will be produced in GM plants around the world and used as building blocks for assembly elsewhere' (*Financial Times*, 21 August 1981).

Such a strategy produces not only corporate reorganization within the major firms themselves, it also forces competitors to reorganize their own operations in an effort to remain viable. For those unable to rival the size of the largest trans-nationals but which operate in the same market, collaborative ventures provide an increasingly important option in the attempt to achieve survival. Thus, in the North-West, Leyland Vehicles, unable to respond to the world truck concept by internationalizing its own operations (beyond already existing activities in India and Nigeria), has been drawn into collaboration with Cummins, the American engine manufacturer, in an effort to cut manufacturing costs. Leyland has also outsourced its gearbox manufacture from Albion (Scotland) to ZF Transmissions in Germany and is being increasingly driven to a position in which, by outsourcing component making, it is becoming an 'assembly only' operation. Once again, while corporate survival may be achieved, the knock-on effects of these organizational changes may well be considerable in the region, both for elements of the Leyland blue-collar labour force and for traditional sub-contract linkages.

A further feature arising from the increasing internationalization of capital is the impact this is having *in situ* for the workforce which does remain in the region. Increasingly, internationalization allows companies to compare production costs at different locations around the globe (Urry 1981) and to develop comparative performance tables between plants. In the North-West, this has produced a contemporary issue where Ford's Halewood plant is consistently being compared with its twin in West Germany – Ford-Saarlouis, which also produces the Ford Escort (see *The Sunday Times*, 20 March 1983). Comparison of production costs at Halewood with those of Ford-Saarlouis and competitors such as Toyota have resulted in 1363 voluntary redundancies requested in January 1983 and increased pressure to change working practices on the shopfloor. As *The Guardian* (21 March 1983) puts it, 'The real subject for negotiation now is how the remaining workforce will be made to perform.'

The process of internationalization has, thus, served to increase the vulnerability of workers in particular localities to closure threats and, at the same time, can be utilized to generate pressure on the workforce by changing working practices in conformity with competitive sister plants. Co-ordinated attempts to reduce wage costs in this way are, however, rarely being conducted with directly comparable measurements of relative performance standardized for variations in accounting systems, levels of technology and of capital investment and seldom placed in the context of differences in internal social organization (see Hodgson 1982).

While, as Dicken (1983) points out, overseas investment and production by UK firms is by no means a new development and the effective measurement of the process is fraught with difficulties, there is clear evidence that in a variety of ways internationalization as a phenomenon is on the increase and that events in a region like the North-West cannot be fully understood without being set in that *global context* within which its prime mover firms articulate their operations. It is too simplistic to see the process of internationalization as some form of 'knee jerk' reaction to cheap labour in the Third World – it does not necessarily imply that the region is being abandoned completely by capital. Acquisition and merger strategies both for production and access to markets, and new technologies, joint venturing and equity investment in the advanced economies all play their part in a process within which direct foreign investment is but one element in a complex package.

TECHNOLOGICAL CHANGE: THE ISSUE OF TRANSIENCY

One of the problems to emerge from a case-by-case analysis of firms in the North-West at a time of deep recession and massive corporate reorganization is that it is often extremely difficult to distinguish between those events which result from a simple purging of excess capacity and those contingent upon more fundamental forms of corporate restructuring. The distinction is fundamental, however, since it conditions the likely response of the regional 'economy' to an upturn. While unutilized capacity remaining in place can, in theory, be brought back into operation, that which has been irrevocably dismantled, sold or relocated as part of a corporate reconstruction programme represents a more fundamental loss to the region, which the arrival of an upturn will not necessarily see replaced *in situ*.

Similarly, technical innovation which has a pervasive impact is proving difficult to isolate. Investment in new forms of technology is proceeding apace and one clear feature of recent industrial change in the North-West has been an observed *increase in the speed of technological change in ways which accelerate the turnover time of the capital stock and, at the same time, both revalue and lower the demand for certain occupational skills*. This produces a condition which, for the sake of shorthand, we have labelled 'transiency'. Under such conditions of transiency, it becomes possible, for example, for rapid changes to take place in the attributes of the capital stock and in the size and occupational composition of the local workforce. By no means all of the employment change is derived, as Massey and Meegan (1982) point out, from such high profile and easily measurable events as compulsory redundancy and closure. The initial effects are likely to be absorbed within relatively buoyant periods by movements within the natural wastage process, rendering the drift of change within the plant or firm largely invisible in the context of observed aggregate shifts in employment. However, in recession, with the pro-cyclical tendency of voluntary separations, natural wastage manipulation becomes less effective as a means of adjusting labour demand and occupational profiles to new technological requirements. Under these circumstances, therefore, the observed level of redundancy and lay-off will increasingly carry with it not only those complex causes referred to earlier but also a proportion of job loss directly attributable to the effects of new technology.

The recent experiences of Ferranti provide a useful case example upon which to explore both the pervasiveness and the complexity of the transiency issue. A regional company with a chequered recent history of NEB acquisition and subsequent re-privatization, Ferranti employs 17,000 workers in 40 plants distributed throughout the North-West, Southern England, South Wales and Scotland. The bulk of the electronics division is located in the North-West, together with a large proportion of Ferranti Engineering and Computer Systems Division. A decision to invest in semi-conductor manufacture at Oldham in the mid-1970s originally arose partly against a requirement for those dexterity skills claimed to be available among women cotton workers and partly as a product of the special incentives being offered to the company by Oldham Metropolitan District. Shortly after the decision was made to invest in Oldham, however, the company outsourced semi-conductor assembly from the North-West to Hong Kong and Korea. Now, CAD technology installed in Oldham allows design to be conducted in the North-West, with digital information transmitted by computer to service the Hong Kong engineering operation. The arrangement of Ferranti's assembly operations is under constant review with a close evaluation of relative labour costs between South-East Asian locations. Currently, Ferranti are said to be evaluating the possibility of returning semi-conductor assembly to Greater Manchester, since a new round of technology demands a smaller input of labour to the process.

Exploring the transiency issue in the case of less exotic technology – Metal Box introduced two-piece canning into its North-West factories in the late 1970s. This achieved a 20 per cent saving for the company in raw material costs (*Financial Times*, 6 August 1980). the process is capital intensive, requires a smaller workforce and demands a seven-day week, four-shift manning system which the trades unions resisted until 1980. More recently, a new round of two-piece canning technology has been introduced, with a £75 million investment programme focused on a new plant at Carlisle. In 1980, while in the process of introducing two-piece canning, the company closed its 1969 vintage Winsford factory. New technology within the 'two-piece' field itself has seen the closure of two production lines at

Westhoughton which had first become operational only in 1980. Prais (1981) examined productivity in the metal can industry in Britain, Germany and the USA and concluded that the slowness of Metal Box to introduce the two-piece can led to the emergence of a large productivity differential between Britain and the USA. Prais suggested that, in the face of such a gap, 'the typical British plant needs to set itself to a halving of employment, while increasing total output by 50 per cent' (1981, 257).

In the field of food processing, Heinz began a major investment project to increase ravioli capacity at their Standish factory in 1974. Twenty-four hour shift working was introduced at their nearby Kitt Green plant soon after, in 1978, making use of high-speed, computer-controlled lines for ketchup, beans and pasta as well as for can-making. Subsequently, the decision was then made to scrap capacity at Standish and transfer production to Kitt Green and this resulted in the final closure of the Standish plant in July 1981.

These three examples indicate for different industry groups the speed of the turnover time which, in a world of fast-moving process technology, can be experienced by key plants in a regional or local economy. It is by no means coincidental that all three companies cited are or have been the recipients of large amounts of state aid of one form or another. It is precisely such assembly and production line processing operations that have been the chief sources of that mobile investment which has responded to area incentive schemes. It is also revealing that, in all three cases, the new technology has been associated with intensification in the use of labour as part of a corporate package which simultaneously involved rationalization and new investment. As yet, the occupational effects of such changes, producing job transiency in a local community, remain to be researched in a North-West context but the work of Green, Coombs and Holroyd (1980), in their detailed study of Tameside, points the way to some of the broad changes that a generation of new technology is likely to usher in. However, whatever the nature and scale of job loss, the evidence so far suggests that replacement investment in new technologies which are likely to *create* employment are unlikely to be forthcoming in the North-West as is the case with the assisted areas in general (see, for example, Smith 1982; Broadbent and Meegan 1982; Cooke, Morgan and Jackson 1984). Indeed, ICL is the classic example of the current speed of transiency in crisis conditions. Opened in

1979, the printed circuit board plant at Plymouth Grove, Manchester, was closed in April 1982, resulting in 312 redundancies.

Clearly, the thrust of new forms of technology, not only in production and in general information transmission but also as a device for facilitating effective management and control from spatially centralized headquarters locations, is having a powerful influence on the nature of job generation and degeneration in regions like the North-West. Under current conditions, replacement investment is being called into use not simply by virtue of its intrinsic merit but by the technological imperatives of competitive survival in a world of deep recession. Set against a context of shrinking rather than expanding demand, its impact on peripheral regional economies is the more sharply felt since it accompanies capacity cutting and corporate restructuring. While it is possible to argue that, in national terms, the dynamic thrust towards an industrial base of high technology can, in the aggregate, be beneficial, an accounting framework at the level of the UK as a whole can fail adequately to recognize the dramatic impact which such a process can have upon already depressed regions like the North-West and upon particular occupational and population groups.

The net amount of replacement capital, while continuing to expand, is, for a variety of reasons, falling short of that required to maintain adequate employment capacity. An important element in this is the propensity of the newest rounds of investment themselves to embody a significant shift in the ratio of employment to output. While such a rise in productivity may be regarded as the salutary effect of investment in 'state of the art' equipment, the new technology itself (as the case of Ferranti showed) offers new capabilities for a spatial division of labour and, in the absence of generally rising demand, may have a depressive effect on regional employment potential even at full capacity of the new capital stock.

The condition of transiency adds to this process a substantial measure of volatility and uncertainty as the 'half-life' of new rounds of investment may substantially be shortened in key sectors – promoting relatively frequent re-evaluations of the investment option as 'old' equipment is written off in favour of new. Once again, new capabilities contingent on new forms of technology against a background of global investment strategy and a new international division of labour will render predictions about regional futures highly uncertain, with the uncertainty itself forming a criterion in the decision-making process.

MONETARISM AND PUBLIC SECTOR MANUFACTURING

While the record of massive rationalization in the North-West continues to grow, particularly within the textiles, automobile and paper manufacturing sectors, a firm-based perspective allows an evaluation of the respective roles of private and public sector owned companies against the general background we have described. This serves to reveal a feature of contemporary industrial change which, while emerging as rationalization, has its origins not simply as an employers' strategy in response to market forces alone. This is the run-down of major public regional companies due to a *politically motivated restructuring of the public sector.* Those companies principally involved are British Rail, Leyland Vehicles Ltd and the British Steel Corporation. Here it appears that the *scale* and *speed* of job loss has been closely connected with the present government's determination to restrain nationalized industries and to re-privatize their activities – a process where the observed outcomes are often more clearly conditioned by the relative strengths of trade union and management positions than by the objectives themselves.

The experiences of British Rail Engineering at Horwich are instructive in this respect, as are those of Leyland Vehicles. During the 1970s Horwich had been chiefly concerned with the repair and overhaul of electric multiple units, wagons and service vehicles. British Rail Engineering advanced proposals in April 1982 to close three such workshops, including Horwich, by the end of 1982. Some 1660 jobs were threatened overall. Later, and in response to a massive local and national campaign by the Horwich and Shildon (Durham) workforce, it was agreed that no workshop closures would take place but that 6000 voluntary redundancies would be achieved at all 13 workshops. The call upon Horwich was for 584 voluntary redundancies, the majority of these now filled by semi-skilled and salaried staff. Under local labour market conditions, however, far fewer skilled workers than anticipated came forward and this resulted, in November 1982, in a further local management call for 242 redundancies, and in February 1983 in a fresh announcement from BR that the Horwich, Shildon and Temple Mills (London) engineering workshops would close with a further loss of 3600 jobs. Horwich was to be closed by the end of 1983 with the exception of its foundry and spring shop, whilst the other workshops were planned to close in 1984. In the context of the Serpell Report (HMSO 1983a), such a move has been castigated by the trades unions as 'closure by stealth' against a political desire to privatize British Rail's remaining engineering workshops.

The similarity between events at Horwich and those at Leyland Vehicles is striking. The restructuring of the commercial vehicles arm of BL has been undertaken against a broader political strategy aimed at reducing the size of BL and of breaking it down into self-sufficient profit centres. In the case of Leyland Vehicles, rationalization has been associated with a concentration of capacity in a smaller number of plants followed by successive reductions in capacity at the three sites identified by Leyland management as the focus for commercial vehicle manufacture in the 1980s. Thus, in November 1981, the company announced 4100 redundancies, of which 1855 were to be at the Leyland plant itself. The aim was by no means simply to reduce capacity (see Shutt and Lloyd 1985). Indeed, assembly capacity at the Leyland site was still being increased with the completion of the new assembly hall in 1981. In effect, many redundancies have been dedicated toward achieving a reduction of in-house manufacturing and have been accompanied by an increase in subcontracting and outsourcing, sometimes accompanied by management 'buy-outs'. Once again, in the case of Leyland, the *scale* of the shift to assembly-only production has been a political compromise, with the company unable to rid itself of in-house manufacturing as sharply, perhaps, as it would have liked. In both cases the short-term outcome has been a growth of *uncertainty* both by management and unions as to the likely future course of events, combined with the subtle process of exclusion as new young trainees are prevented from participating in a stagnant labour market. Within this group of firms, therefore, changing political views upon the role of publicly owned corporations have been confronted by concerted union action to deflect and, in the short-term, slow down the process of rapid restructuring.

CORPORATE FUTURES IN THE NORTH-WEST

Against a background such as the one outlined above, it is, perhaps, not unreasonable to expect that the processes we have identified will persist in the coming decade. The processes of concentration and centralization will continue and such major titular changes as take place among the region's prime mover companies are likely, as in the recent past, to reflect merger and acquisition rather than outright liquidation or the emergence of new corporate giants by internal growth. Further, often dramatic, changes are to be anticipated in the nature of production which will generate a continuing decline both in the quantity and quality of available employment.

Some of these changes are already in progress as long-standing corporate restructuring programmes continue to unfold at companies like Courtaulds, Tootal, Renold and ICL. Other changes are, as yet, only embryonic as key regional companies begin for the first time to restructure in response to pressures of shifting markets, new technologies, aggressive competition, over-capacity and changing political priorities within the UK.

As we have shown, it is extremely difficult, under contemporary conditions, adequately to disentangle the motive forces for change in what is, in essence, a complex and fully-connected set of processes. For the bulk of the region's key companies, pressures toward internationalization, the rationalization of redundant capacity and the drive to compete in the application of new technology are exerting a *simultaneous* influence.

While causes are complex, outcomes are revealed in the North-West as rationalizations of manufacturing capacity involving both the expulsion of workers by redundancy and their exclusion by non-recruitment. While the former is readily quantifiable, the latter is a more insidious process – denying young would-be participants, in particular, an opportunity to join the world of employment. Even the region's previously most stable sectors have, in recent months, begun to reveal disturbing patterns of change which cannot augur well for the future.

With a sound resource base, a long history in the region and a recent record of massive, generally subsidized, investment, the heavy chemicals sector is beginning, for the first time, to see a sharp rise in redundancies after a decade in which manpower levels had remained relatively constant. For ICI's mixed chemicals operations, energy costs, the demise of linked industries in the region (Courtaulds and Bowater), new technologies and a fresh drive to achieve economies of scale occur at the very time when a new round of investment has favoured the development of a £250 million investment in PVC and chlorine manufacture at Wilhelmshaven in West Germany. Within the North-West, restructuring against a new corporate plan and capacity problems led to a rationalization programme in 1979 and a call for 3900 redundancies in 1981. Local reactions are aptly summarized by the *Financial Times* (31 July 1981) in the following terms: 'Mond has been insulated from industrial troubles for so long that the change came as a shock to workers in the North-West whose families expected automatically to find employment in the Runcorn works.' For an industry of such key importance to the regional economy (Cheshire County Council 1982) and of such traditional stability, these emerging trends are particularly disturbing.

In parallel with these events at ICI Mond, another of the region's capital intensive growth sectors has been experiencing difficulties. Overcapacity in oil refining which has seen the capacity utilization of European plants fall as low as 60 per cent (IFCEGWU 1982) saw the closure of the Burmah refinery at Ellesmere Port in 1982 and has provoked a major redundancy programme at the Shell-Carrington petrochemicals complex.

Looking at other traditional growth (or at least stability) sectors in the North-West, incipient trends are little more encouraging. In the region's strong aerospace sector, for example, rationalization at Rolls Royce, Barnoldswick and Lucas Aerospace in Burnley is contingent upon the falling general demand for civil airliners and the abandonment of the Lockheed Tristar programme with its Rolls Royce RB 211 lead engine. Output at Barnoldswick fell, for example, by 30 per cent between 1981 and 1982 with further reductions in 1983. In general terms, only the military aircraft programme remains buoyant at British Aerospace, Preston. The region's other aircraft plants and their associated suppliers seem vulnerable in the short term to the decline in the civil market.

Within other high technology fields in the North-West, there exists a group of traditionally successful companies sharing a common characteristic in their high degree of dependence on public sector contracts in the power generation, telecommunications and traction fields. These, too, for both the general reasons we have outlined and for more specific ones contingent upon their close ties to the public sector, have been forced to implement major rationalization programmes in recent years.

GEC and Plessey form the core of the group. Both have a history of innovation and high technology linkage, employing workforces generally attuned to the ramifications of technical change but more normally against a *background of growth and expansion*. Under modern circumstances, such companies find themselves forced to confront the cutting edge of new technologies (particularly in the CAD/CAM field and the electronic office) in the context of falling general demand and the destabilizing thrust of government policies to liberalize public sector purchasing policies.

In the final analysis, therefore, even for the strongest regional companies in the most advanced and secure sectors, the impact of corporate restructuring, internationalization and the introduction of new technology will serve to depress levels of future regional employment even at the full capacity of capital stock. As we pointed out earlier, it is vital to recognize that replacement investment, at a time of volatile changes both in corporate organization and in new technology, will rarely be by a capital good of identical character. In the dynamic process of change, which we would see as *revolutionary* rather than evolutionary, regional futures would be anticipated to contain fundamental changes both in labour requirements and in the flows of raw materials and intermediate goods contingent to the production process. Far from being a product of simple overcapacity in recession, the events and processes that we have described for the North-West represent a fundamental structural change which national recovery from recession will by no means redress.

Conclusion: methodology and policy

Evidence from the North-West shows that under contemporary conditions both the *scale* and the *spatial incidence* of employment decline have changed in dramatic fashion. The basic economic structure of the North-West is being radically altered as manufacturing capital restructures its operations. Moreover, this time round, the regional economy is not only subject to manufacturing but also to services employment decline. Public sector employment cuts, corporate restructuring in the banking, finance and distribution sectors and a new round of technology in the office are offsetting the traditional 'lifebelt' effect which service employment growth has offered to a region with long-standing losses in its manufacturing base.

Significant changes have taken place in the spatial incidence of these employment losses. In particular, recession in a peripheral region has been associated with a more widespread incidence of job loss. The employment problem of the inner city which attracted the attention both of government and the media in the mid-1970s has seen its clarity of definition as a 'spatially localized' problem area increasingly dulled. While unemployment and job loss in the urban cores have themselves continued unabated, the syndrome has spread more widely to erode the differences between inner city and suburbs and free-standing towns. In particular, those suburban industrial estates which represented the employment growth points in the 1960s and early 1970s have now become the setting for major corporate redundancy programmes among branch plants opened during the years of relative success. Even those more successful free-standing towns which were acknowledged by general consensus to be the buoyant growth points of the nation have (at least in the North-West) become afflicted with job losses as their dominant companies have restructured and confronted recession. In response to the observed dispersion of job loss and the increasing pleas of the outer city local authorities for assistance, the government's more tightly drawn map of regional assistance seems increasingly irrelevant – reflecting the fundamental inability of fixed areal definitions of areas of 'intractable unemployment' adequately to reflect the dynamics of employment change in a period like the present. Against a background of the international reorganization of capital at a time of rapid technological change, it is becoming increasingly obvious that continuing with the traditional form of regional policy on a slimmed down areal basis combined with a host of inadequate and *ad hoc* responses which simply encourage firm relocation is wholly inadequate.

Given the depth of the economic crisis, the inadequacy of the government's policy and the marginal impact of many local authority economic development assistance schemes, it is not surprising that regional and urban policy issues have returned to take a more prominent place on the national political agenda. While the more radical initiatives of the Labour and Alliance parties are now pre-empted by the result of the June 1983 election, regional inequalities can only deteriorate and the pressures for change remain. In view of the flourish of criticisms from outside and within its own ranks, the government was forced to announce a major review of government regional policy in July 1983 and to publish a White Paper later that year (HMSO 1983b).

In considering the future form of regional policy, it is as well to be aware that research methodologies influence policy prescriptions. From the perspective of this study, the examination of the region's key corporate enterprises shows that the processes of change comprise far more than localized responses to adverse factor-cost comparisons at a time of shrinking markets. Approaches which identify job loss and weak economic performance primarily by reference to those features with which outcomes are *associated in the regional context* (old vintage capital, unfavourable industry-mix, poor infrastructure) must now be discarded in favour of a method of analysis which attempts to identify causal relations which are primarily *external to the region*. Whilst it is not denied that regional factor-costs and the general infrastructural setting will influence capital, these conditions are by no means sufficient to explain the nature of observed events and processes.

In the specifically regional sphere, our method of analysis shows how a region like the North-West has been drawn more tightly into the global production networks of those dominant trans-national companies whose local activities underpin the regional manufacturing economy. Under these conditions, the logic which integrates the regional space economy is no longer physical propinquity among activities sharing the benefits of some clearly defined regional resource or labour pool. This has given way to a new logic of integration based upon the global circulation networks of those prime mover enterprises which maintain capital stock in the region. In response to this, the spatial-hierarchical division of labour evolving from the current reorganization of production processes draws the region more tightly into the new international division of labour, and it is against this background that the recent past and likely future performance of regions like the North-West must realistically be evaluated. *Explanations of events and, therefore, the formulation of policy initiatives which seek to exercise an influence upon the mechanisms of change must confront both a changed world and an evolving understanding of process.*

Damesick (1982), in calling for a review of policy, has suggested that three broad options present themselves for discussion. The first, put simply, is to refuse to recognize any specifically regional dimension to economic policy making. This is closely in accordance with the general shift of emphasis in government policy towards giving top priority to the battle against inflation and keeping a tight hold on public expenditure. It is only when the national economy itself turns round that a reduction in regional inequalities can be achieved, and to help this process many would advocate the abolition of regional aid altogether in order to cut public expenditure. This climate of opinion can easily be buttressed by emerging views from across the Atlantic that policies of 'planned shrinkage' need seriously to be considered rather than active policies for regeneration. Kennett argues, for instance, that

> the encouragement of population and industry to places which have been rejected by entrepreneurs for decades seems . . . to make little sense. A policy which allows the orderly and balanced declines of such locations by planned out-movement of the unskilled and unemployed is advocated. (Kennett 1982, 30)

Kennett wonders, for example, whether, given limited national resources, we can afford to maintain two conurbations in the North-West. Such an approach, of course, is far from new. Its roots are to be found in the mid-1930s debates on regional policy. A second dimension of the debate identified by Damesick is that relating to the scope for keeping the *status quo*. That is to reaffirm the case for continued regional and sectoral preferential assistance, whether on grounds of social equity or the apparent belief that existing policies offer some relief and means of avoiding alienation, local unrest and the disruption of the state. Advocates of the market forces approach may still see a role for traditional regional policy as a minimum 'safety net' designed to ameliorate extremes of deprivation and thereby defuse the potential for any more disruptive form of social action. More liberal opinion would perhaps interpret them as policies generally conceived to promote greater equity or to provide a prescription for the renewal of 'self-sustaining growth' in depressed areas, placing the need to adjust and adapt to existing economic circumstances at the fore of policy objectives. To argue for a renewed approach to regional problems from the policy perspective of the 1960s it is necessary to state how new initiatives such as those advocated by the recent report of the Regional Studies Association (RSA 1983) will have a greater expectation of success than those previously attempted.

The third dimension of the debate is that relating to the scope for *locally based* initiatives in the face of emerging economic trends. Whether articulated by those who see no role for regional policy or those who want to see present policies retained, it is usually conceived within the self-help framework and emphasizes the role of local authorities, venture trusts and local enterprise agencies in 'pump priming' market forces in their particular areas. Central to this approach is the belief that local rather than central authorities are best suited to respond sensitively to local business needs and that small firms and new firm formation will provide a major solution to the regional employment problem. This approach, of course, ignores the reality of linkages between large and small firms and the centrality of large organizations in employment creation. Another body of work conducted in the North-West (see Lloyd and Dicken 1982) shows the volatility and vulnerability of small firms operating in the urban environment and highlights the limitations for a policy based upon indigenous enterprise. Under conditions where 60 per cent of such firms fail to survive three years and where the mean plant employment is eight workers, little job replacement commensurate with need can be expected. Indeed, the existence of market constraints and the essential nature of the links between small and large enterprise render the small firm option little more than palliative.

From the perspective of the research reported here, it is necessary to stand back from the nuts and bolts of the policy prescription debate in order to reconsider the realities revealed in the case study of the North-West. First, as we have suggested, a decade of recession has promoted an unprecedented collapse of the regional manufacturing base. Second, the continuing scale of job loss is of such an order and so generally widespread that both traditional and more recent Conservative policies are both marginal in scale and largely internally redistributive in their impact. Third, the continuing centralization of capital and the internationalization of production provide little evidence so far that the existence of a growing regional labour reserve will provide an attraction to capital even given a cyclical upturn. Moreover, if and when the upturn takes place, changing technology alone is likely continually to increase the scale of unemployment. Whilst wage reductions are clearly considered in some quarters as the only viable step in rendering surplus labour more attractive to internationally mobile capital, the scale of existing wage differentials in a world configured by the new international division of labour is already such that this would virtually be impossible to achieve without major social disruption.

It is, therefore, important to consider what the objectives of regional policy should be, faced with the situation we have revealed for the North-West. A primary concern to reduce spatial inequalities in a world configured according to the necessary divisions of labour and social relations of capitalism is unlikely seriously to be addressed by any policy which takes the capitalist mode of production as given and seeks to leave it unchallenged.

From the perspective of labour in the regions, however, where the objective is to provide benefits to workers both in and out of work, the prime general requirement is for massive investment of the kind capable of generating sharp increases in labour demand. Clearly, policies which depend on 'leverage' and providing incentives for mobile capital are inadequate to achieve this. Inevitably, the solution to regional problems lies in greater public expenditure and increasing control of the investment process in order to both raise overall expenditure and to direct it to more socially important activities. Investment in railway electrification, education, health and infrastructure provision can provide the necessary impetus for other than a consumer-led boom based on expanding imports. Such a policy must be founded on an alternative economic strategy for the nation as a whole – one which recognizes that recourse to market forces is both an antiquated ideal with uncertain outcome and one bound to increase regional inequalities.

The precise direction of an alternative economic strategy is, of course, the subject of much greater debate (see, for example, Aaronovitch 1981; Glyn and Harrison 1980; CEPR 1981). At the specifically regional level, given a flow of investment upon which to work, there is much to be done in monitoring and redirecting the spatial outcomes of expenditure in accordance with agreed objectives. At the present time, much explicit regional or urban policy is, as we have seen in the case of, say, the Horwich closure, overtaken by the spatially distributed effects of other actions of government. With the use of a wider accounting frame than cash limits, regional objectives could be achieved (say through the PESC system) more effectively, in ways which traditional forms of regional policy have failed to address.

It is necessary, in particular, to identify and mould those unintentional outcomes of policy which have specific regional effects and to develop more consistent strategies for national industrial investment. Co-ordination with regional objectives should be an implicit element of national policy *at the outset*.

For the development of a national industrial strategy co-ordinated with regional policy, the TUC (1982) has argued that a new centre for economic planning within government is required, and proposes that Regional Development Planning Authorities (RDPAs) should be established in the English regions to match the functions of the Scottish and Welsh Development Agencies. Whilst there is little doubt about the need to bring greater co-ordination to the variety of contemporary *ad hoc* arrangements, and the problems of zero-sum promotional games have been alluded to earlier, the creation of another tier of government would be contentious. The North-West 'region' which forms the subject matter of this chapter would be better conceived of as a cluster of labour markets whose boundaries are not fixed but are ever-changing and which, in general terms, reflect the disadvantaged position of industrial workers and those attached to certain formerly labour-intensive services. Concepts of regions based on arbitrary aggregate social and economic indicators do not provide the most appropriate units from which to develop alternative strategies, and it may be that the county level is the more useful administrative basis from which to mount a more appropriate strategy derived from national planning.

Clearly, however, institutional arrangements will be of little concrete value unless greater direct control and guidance of the investment process can be achieved both within financial institutions and large corporations. In the North-West the Stone-Platt case indicates something of the current friction which exists between industrial and financial capital in the UK and lends support to the view expressed by Minns (1982) and others that there is a need for greater control of financial institutions in order to achieve more positive investment in production. In particular, evidence from this case study appears to indicate that the demand for more rational and longer-term responses to current events is by no means solely the preserve of academics and radicals:

> If we take textile machinery as an example, the range of PSL machines was perhaps no longer the world leader, but they were robust and well proven, they had a long-established reputation for quality and important developments were in hand to re-establish their excellence. The research and development costs needed to bring a new friction spinner into production, to add automated features to other models, and to reduce manufacturing costs by improved design or value engineering would not have exceeded £5 millions over a 2 year crash programme. Is such a grant too much to ask of a great textile manufacturing country, that could still be a world leader in textile machinery and mill technology? Or should we write it all off, including the skilled people involved, as being something that is not of sufficiently advanced technology to merit our assistance? . . . In Stone-Platt's case, we had good products, good labour relations, world markets and skills, and a practical plan for survival – and yet the 'rug was pulled' when we were more than halfway through the rescue job. On behalf of a large number of people, I submit that this is not good enough. ('The chairman on the receiving end', *Guardian Financial Extra*, 26 April 1982)

We would argue, therefore, that the creation of a network of county Enterprise Boards to channel institutions' monies and intervene in the restructuring and development of industry along the lines of the West Midlands and Greater London Enterprise Boards should be seen as a welcome positive initiative in local economic planning. It is, however, important to place these experiments in perspective. Without a change in national policies which aim to achieve fundamental changes in the private accumulation process, local initiatives, even those which take a more interventionist stance, cannot be relied upon to eliminate regional inequalities. Indeed, with a stronger network of regional planning boards and enterprise boards in current conditions, the danger is that the zero-sum game in which local interests vie with each other to offer inducements to private capital would simply be played out on a more intensive scale.

Notes

1 The notion of capital 'age' can, however, be misleading if applied too literally. 'State of the art' considerations may make newly-installed plant 'age' prematurely in such fields as computing (see Duncan 1982) or others with fast-moving technology.

2 Key firms in the service sector were not omitted for reasons of choice but simply through the absence of equivalent data.

3 The nature of available data at establishment level makes the use of manual worker employment and manufacturing sector firms an involuntary choice.

4 The data source (Factory Inspectorate records) has an inevitable tendency to underestimate *in situ* change.

5 Lowry (1982) reports: 'The effect of the acquisition and the decline in the UK has been that Pilkington can no longer be regarded as a UK company with substantial overseas interests, but rather as a European company whose largest market is in the UK.'

3
Editorial introduction

Townsend and Peck agree with Lloyd and Shutt in rejecting attempts to get at causality itself through extensive analyses of aggregate statistics of industrial change, whether these be aggregated by region or sub-region or by industrial sector. They, too, argue strongly for the importance of studying individual, named, corporations. On this last issue, their chapter is particularly useful in addressing some of the problems, both practical and conceptual, which face the researcher who sets out on this kind of analysis. But, again, like the authors of chapter 2, these authors do not wish their analyses of corporations to exist in a vacuum, and they therefore devote an early section of their discussion to the wider economic context within which the corporations are operating. These two levels of analysis also highlight a range of issues which again recalls some of those pointed to by Lloyd and Shutt. At the wider level, the facts of internationalization, of political strategies by governments, and of conflict between capital and labour are inescapable. At the level of the corporation the importance is stated of going beyond figures for employment to look at output, and at the rate of profit (a role which is certainly not mechanistic), at the importance of technology, at the form of social relations through which the mechanisms of change are played out, and vitally, at non-market influences.

But in spite of these apparent similarities, there are fundamental differences of approach between this paper and that of chapter 2. Perhaps most clearly, there are differences in the analysis of the corporations themselves. Two of these are of particular importance to the debate in this collection. The first concerns the status of the corporation as an agency of change. Like Lloyd and Shutt, Townsend and Peck argue that one reason for focusing on major corporations is quite simply their empirical importance in the actual process of employment change in Britain at the moment. But they also strongly stress another reason: 'The case for analysing larger redundancies primarily by corporation does not, however, simply rest on this level of dominance in generating recent dramatic changes in employment. It is also based on the importance of corporations' financial strategies as a direct or proximate origin of redundancies.' The authors, in other words, clearly stress the importance of corporations as agencies of change. The second difference of this analysis of corporations from that in chapter 2 concerns what is being looked for. The authors are clear that 'this is not a good field for the making of deterministic laws'. And yet they are also determined that their analyses will not get caught in what might turn out to be the idiosyncrasies of an individual corporation, for corporations, as both they and Lloyd and Shutt clearly recognize, vary greatly in their behaviour both between one another and over time. How is this problem to be solved? Much depends – as usual! – on the way in which the problem is conceptualized. Townsend and Peck see it as a problem of representativeness. They wish to draw conclusions which, though perhaps not 'deterministic' in the strong sense, are none the less empirically general, and possibly empirically generalizable. They therefore take up explicitly the thorny issue of how many corporations to study, and how to select them. They also derive hypotheses about a range of factors which may influence the behaviour of corporations in producing particular geographies of closure. This, then, is different from the approach of Lloyd and Shutt. Here the emphasis is on identifying important common factors rather than on unearthing underlying processes.

This difference between the two approaches is related to others. Not only is the analysis of corporations different, so also is the analysis of the 'wider sphere' and of the relationship between the two levels. For Townsend and Peck the wider context is not so much the underlying forces of international capitalism as the empirical context within which the analysis of corporations must be set. Their 'wider economic context' therefore highlights the important aggregate changes which are both backcloth to and result of corporative behaviour. The relation between context and corporation is therefore one of context and agency, rather than of underlying forces and agency. For Townsend and Peck the agencies of change are thus of prime importance.

This position is followed through with great rigour in the discussion of policy. In the early part of the paper the basic position is clearly established: 'If, as implied above, these corporations are the driving force behind major changes in manufacturing employment levels, then policy formulation . . . must be based on a consideration of their strategies and activities.' It is the agents of change which must themselves be intervened in. This focus on agents is also carried through into the policy-making arena itself, where the different departments and bodies responsible for policy are identified. What Townsend and Peck's analysis makes clear is that policies for the geography of employment are designed and promulgated within a *political* arena.

The policies themselves clearly begin from the logic of the analysis. The authors call strongly for structural intervention in individual cases, in individual corporations. They also discuss the possibility, now increasingly a reality, of, in a sense, changing the personnel of agency, through management or worker buyouts. In this they are taking seriously the room for manoeuvre offered by their own recognition (and that of most others in this collection) of the existence of what might be called a 'behavioural' element in the determination of redundancies. But, finally, in following through the logic of their policy recommendations, the authors also recognize the existence of those wider underlying forces – the rule of profit, the internationalization of production – which lie beyond and above, as well as within, the individual corporations. Their proposals also recognize the need, therefore, to withdraw production on occasion from the play of profit and market forces – from, that is, the dominance of capitalist relations of production.

3

ALAN TOWNSEND &
FRANCIS PECK

An approach to the analysis of redundancies in the UK (post-1976): some methodological problems and policy implications

Introduction

Any review will show that two different approaches have been used to analyse recent changes in the patterns of manufacturing employment in the UK. One approach has made extensive use of aggregate statistics to describe the patterns of change across regional and *sub-regional* boundaries (Fothergill and Gudgin 1979, 1982; Keeble 1976, 1980; Martin 1982; Townsend 1980a, 1983a). A second approach, while stressing the importance of large industrial companies in shaping employment patterns, focuses analysis on particular *industries* (defined by their category of product; Massey and Meegan 1978, 1979, 1982). The former approach may provide useful descriptions of net geographical changes, and the latter may afford valuable conceptual insight into the mechanisms of some of these changes. However, we suggest that the most effective way of identifying the origins of larger redundancies and of drawing out policy implications, is by reference to the activities of the *named corporation* across a national set of regions or sub-regions. This paper explores the benefits and the problems associated with this approach. It refers to

Table 3.1 Leading job losses in the 'peripheral regions' of the UK (the largest cases reported in each area, October 1976 to October 1981)

Sub-region	*Parent organization*	*Jobs lost*	*Rank*
1 N. Ireland	Courtaulds	3,300	4
2 Merseyside	British Leyland	4,600	N
Scotland			
3 Borders	Cawdaw Industrial Holdings	100	—
4 Central	British Petroleum	250	1
5 Dumfries & Galloway	Spillers	100	22
6 Fife	British Steel Corporation	300	N
7 Grampian	Unilever	400	5
8 Highland	BAT Industries	450	3
9 Lothian	British Leyland	2,100	N
10 Strathclyde	Peugeot	9,000	336
11 Tayside	NCR Corporation	1,550	404
Wales			
12 Clwyd	British Steel Corporation	8,100	N
13 Dyfed	Duport Steel	1,600	302
14 Gwent	British Steel Corporation	8,500	N
15 Gwynedd	Bernard Wardle	400	—
16 Mid-Glamorgan	Hoover	2,200	250
17 Powys	GKN	800	20
18 South Glamorgan	British Steel Corporation	3,300	N
19 West Glamorgan	British Steel Corporation	10,100	N
North			
20 Cleveland	British Steel Corporation	11,300	N
21 Cumbria	British Steel Corporation	3,400	N
22 Durham	British Steel Corporation	4,600	N
23 Northumberland	Lonrho	300	26
24 Tyne and Wear	British Shipbuilders	3,800	N

Source: Survey of all reports of job losses in the *Financial Times*, as in Townsend (1981). Entries refer to the sum total of job losses in the respective areas, summed from all applicable reports, which may refer to different plants, subsidiary companies and dates.

N = Nationalized.

the patterns of redundancies associated in particular with *closures* and *part-closures* of productive capacity, which occurred in the UK after October 1976.

The merits of analysis by corporation

Few would doubt the importance – where it were possible – of considering changes in industry with reference to the financial control exercised by the larger private and public corporations. This is particularly true with regard to major redundancies where at least the memorable large cases tend to be associated with relatively few 'big names'. Table 3.1 shows, for example, the largest redundancies reported in the *Financial Times* in each sub-region in the 'peripheral' regions of the UK during the study period (Northern Ireland, Scotland, Wales, Northern Region and Merseyside). Of the 24 cases, 11 were in nationalized corporations and another 11 in the UK's leading 500 companies (i.e. those ranked by turnover above 500 in the *Times* 1000, 1981–2). A similar picture emerges in the 'normally prosperous' regions and the 'manufacturing heartland' (Townsend 1983a, 130 and 112–13 respectively).

The source used in Table 3.1, the *Financial Times*, covered in all just over half the equivalent GB manufacturing total of redundancies in the best official source (series ES955 of Manpower Services Commission; Townsend 1983b). *The case for analysing larger redundancies primarily by corporation does not, however, simply rest on this level of dominance in generating recent dramatic changes in employment. It is also based on the importance of corporations' financial strategies as a direct or proximate origin of redundancies.* Analysis of particular corporations' decisions, as they affect the individual production units which they control, tackles these mechanisms of change more directly, avoiding many of the assumptions which are inherent in analyses based on sub-regions or defined industries. Two of these assumptions are of particular importance. First, neither of these alternative approaches differentiates adequately between types of production unit. Analysis of sub-regional changes using shift-share analysis (for example, Fothergill and Gudgin 1979) discriminates by industry type only (usually by minimum list headings of the Standard Industrial Classification), involving the implicit assumption that other differences between plants (for example by size, productivity or skill ratio) in the same sub-region are unimportant. Analysis of the type used by Massey and Meegan (1982) avoids this assumption conceptually, but in operational terms it may also rely heavily on assumed regularity in the patterns of change amongst *ad hoc* groups of plants in the same industry and statistical area. Second, these approaches tend to ignore relationships *between* industries, which in many circumstances can be crucial to the overall understanding of job losses. Of prime importance in this respect are the effects of 'conglomerate' corporations which have commercial interests in many activities. Corporation *A*, specializing in industry *X*, may see its plants in industry *Y* in a very different position of relative profitability compared with corporation *B*, which specializes only in industry *Y*. Again, the linkage of a particular plant, in say industry *H*, to a particular firm in industry *J*, may give it a very different production and employment trend from other plants in industry *H*.

Finally, the analysis of redundancies within the major corporations has obvious policy implications. *If, as is implied above, these corporations are the driving force behind major changes in manufacturing employment levels, then policy formulation must be based on a consideration of their strategies and activities*. Use of corporate data

> can have a part to play in research at the national level, particularly in substantiating the spatial behaviour of Holland's (1975) 'mesoeconomic' sector; it would seem practical to have a 'BL column', a 'Courtaulds column', etc. in future national tables of the components of employment change, to strip away the present anonymity of such corporations' impact on the employment geography of Britain. (Townsend 1981)

This reinforces still further the importance of research based on named corporations.

Some preliminary problems

The lack of substantive research into patterns of redundancies in the major corporations can in large measure be attributed to practical difficulties. Sources of information may tend towards the superficial, as is the case with company reports, or else are inevitably unsystematic, as with reports in the press and media in general. Survey methodologies too are problematic, relying on the goodwill of top-level management, and may be limited by confidentiality constraints, staff changes and the sheer problem of locating a manager who had an overall purview of the corporation's set of rationalization decisions, if indeed there was such a manager.

The corporate approach also presents various conceptual difficulties. Despite the abundance of literature concerned with large industrial corporations (variously termed 'enterprises' [Watts 1980] and 'Giant Firms' [Prais 1976]), there is comparatively little consistency in the definition of such organizations. In a recent review, Hayter and Watts (1983) point out the difficulties of defining 'firms' and the boundaries between firms and their 'environment'. *Legal* ownership, for example, is not always a fair reflection of functional or financial *control*. One firm can have a large shareholding in another without being its legal owner; for instance ICI had till recently a 49 per cent shareholding in the textiles firm Carrington Viyella. Such conceptual difficulties are then compounded by others of an operational nature, for it is not always easy to identify ownership, particularly in a period of rapid change as witnessed since 1976.

Further problems arise in designing the plan of research. *Taking the two extreme cases, this involves a choice between scanning a wide range of corporations* (as in Townsend and Peck 1985) *or selecting one or two to analyse in depth* (as in Peck and Townsend 1984). In relation to public policy formualtion, there are undoubted advantages in casting a fairly wide net. However, the effectiveness of such a design in establishing the often complex origins of redundancies is debatable, and it can lead to fairly superficial generalizations. The other extreme, the 'case study' approach, also has its obvious price. The solution to this problem of research design may lie in careful selection of corporations either singly or in groups. Selection can be based most simply on employment levels, as a statement of the potential impact of larger employers. Lloyd and Reeve (1982), for example, direct their attention towards the 54 largest employers in the North-West region (those with over 2500 shopfloor employees). Alternatively, one can select all the major employers within a specified region or set of regions and in one industrial sector (Healey 1981, 1982). Selection on the basis of distinctive financial strategies also has considerable logic, concentrating research for example on corporations which were known to be investing heavily overseas while capacity was cut back in the UK. This approach, however, is probably the one in which evidence is the most difficult to collate. Finally, selection can be based on the characteristics of redundancies in different corporations, focusing, for example, on those where closures are the dominant form. This approach is the one elaborated later in this paper, following the next section which reviews the empirical context of change.

The economic context for the analysis of recent major redundancies

Massey and Meegan (1978) have stressed the importance of linking regional and sub-regional changes with changes taking place in the national and international economy. This is equally important when analysing changes within individual corporations. It is now a commonplace that the period 1976 to 1981 saw mounting changes in the structure and levels of manufacturing employment which have no precedent in post-war Britain. These changes were unique not only in the severity of employment decline, but also in their widespread geographical nature (Townsend 1983a; Martin 1982). Some industries associated with post-war prosperity, notably the vehicle industry (Dunnett 1980; Law 1982), have been particularly hit, while more traditional industries have experienced continued and intensified decline (Townsend 1983a, 50–8, 176–80).

Aggregate employment trends are sharply defined in the latter part of the study period compared with previous years. Employment in manufacturing industry increased slightly from 1976 up to late 1977, reaching a GB total of 7.22 million. From this plateau, the total fell dramatically after mid-1979, reaching 6.06 million in the Census of Employment for September 1981, and 5.44 million three years later (*Employment Gazette*, March 1985). Over the same period, the index of manufacturing production rose from a base year in 1975 (100) up to 104.3 in 1979, only to fall back to 89.4 for 1981 as a whole *with no significant change in 1982* apart from gentle, further decline prior to an eventual increase in 1984–84.

Trends in manufacturing productivity indicate that the employment decline was not simply a direct consequence of falling output levels. Productivity tended to increase throughout the period, only declining a little from mid-1979 to mid-1981, before then showing an accelerated increase. This important pattern can be interpreted in various ways. First, and most obviously, it is possible that there has been the familiar recessional 'shake-out' of 'inefficient' production units across industry as a whole, resulting in overall improvements in productivity after a decrease in output and employment. Alternatively, it may also indicate that recession has hit labour-intensive industries more than others. This fits with high rates of redundancy in the textiles, clothing and footwear industries, but is not entirely consistent with high levels of redundancy in vehicles and iron and steel (Townsend 1983a, 53). Nevertheless, this trend is undoubtedly associated with structural changes of this nature. Third, it is clear that recession has weakened the powers of labour to resist changes in working practices, providing management with leverage to enforce changes which in better economic circumstances might be opposed more effectively. This process can be seen in operation in reported threats of closures and redundancies in the financial press. Just as one example among many, 8000 employees at the Harland and Wolff shipyard in Belfast were told they must 'significantly improve performance if the yard is to avoid massive redundancies' (*Financial Times*, 20 February 1979).

The economic pressures on management in this period certainly give cause to expect such policies. The UK fall in industrial output (1980–2) was followed in 1982 by a decline in world manufacturing production which intensified competition between producers. Gross trading profits, which had risen steadily from 1974 to 1979, fell heavily by 1981. The performance of companies confined to the manufacturing sector was probably much poorer than average. Mention must also be made of a parallel trend in direct investment overseas by UK-based companies. Up to 1980, such investment fluctuated around one billion pounds per annum, but, after the new government's relaxation of exchange controls, it increased sharply to over four billion in 1981 (*Barclays Review*, August 1982). Most of this went to EEC countries, particularly West Germany (25 per cent of direct investment overseas: *Financial Times*, 20 March 1982). Tentatively, this suggests that while UK industry has suffered substantial disinvestment, many companies appear to be moving capital overseas, a process which Taylor and Thrift

(1981) see as a significant contributor to British de-industrialization.

The corporate context

Relating these wider economic trends to the activities of individual corporations is difficult and potentially dangerous, for one cannot assume that aggregate trends in the economy are typical of the experience of parts or the whole of individual corporations. Massey and Meegan (1982) offer some guidance in this respect, with their three *forms* of job loss, 'intensification', 'investment and technical change' and 'rationalization'. They stress that these forms of job loss are closely related and overlap in practice. These categories may therefore be fairly difficult to handle in the corporate context. A corporation, for example, could operate 'intensification' policies in one plant, introduce technical changes in another, while closing a third producing for the same market. Even within the same plant, reductions in capacity could be (and are likely to be!) accompanied by efforts to improve the productivity of the remaining workforce.

None the less, the classification is taken as central to the analysis of job losses at the corporate level. In particular, the approach stresses the need to relate changes in employment levels to changes in output, either at the national level or when dealing with individual markets or corporations. We suggest, therefore, that the analysis of job losses in the corporate context should wherever possible commence by classifying redundancies (or job losses in general) into different types depending on output levels and changes.

Initially at least four categories of job loss can thus be identified conceptually, related to the above categories (and defined in such a way as to avoid assuming causality):

(a) employment loss associated with total output loss;
(b) employment loss associated with proportional output decline;
(c) employment loss associated with static output;
(d) employment loss associated with increased output.

Evidently this classification can be applied to whole economies, complete industries, or individual plants. This creates problems for our analysis, which is primarily concerned with corporations whose production may spread across many industries experiencing different trends in output and employment. Equally, there are problems in relating job losses to output changes at the plant level. One plant may shed employment within the context of an overall increase in output and employment in other plants belonging to the same corporation. These problems can partly be resolved by selecting corporations with distinctive aggregate patterns of employment and output. For example, one could select corporations whose employment and output are both in decline, where it is reasonable to assume that job losses at one plant are only rarely accompanied by compensatory increases elsewhere. The analysis could then concentrate on the plant level. This chapter explores this approach, focusing in particular on plant closures (employment loss associated with total output loss). While in most cases plant closures are clearly defined, on many other occasions 'part-closures' may occur, where one section or production line within a larger factory, or on a site adjacent to other buildings, may be withdrawn from use. Though more difficult to identify, part-closures have much in common with 'full' closures. Indeed, the distinction between the two is rather artificial. It can be affected by accidents of history, for instance whether or not two particular plants – which might have been built on separate sites or in separate towns – were built juxtaposed within the same complex. Further use of the term closure will therefore include 'part-closures', although the problems of identifying such events are recognized.

Despite these difficulties, there is a number of reasons why it makes sense to focus on closures. First, 'full' closures, at least, focus attention on a field of great concern to policy because they can deny an area a share of any renewed upturn in production. Second, national trends in output and employment, as described earlier, suggest that many corporations may be suited to such analysis. This view is partially confirmed by the balance of closures to other redundancies in reports in the *Financial Times* in this period (1976–81). We may also study what proportion of the announcements of job losses drew attention to any new investment anywhere in the respective corporation. In North-West England, for example, only 36 out of 231 did so (15.6 per cent). Included in these figures were 62 cases in Merseyside, where only 10 cases mentioned 'technical changes' of any kind. Most of those that did so pre-dated the deep recessionary period after mid-1979. Third, although information about levels of output within individual plants is generally difficult to find, closures are usually fairly clearly visible, with the possible exception of part-closures as suggested above.

The analysis of the *geography* of job loss (related here to closures) involves the following question: *why were some plants within particular corporations closed and not others*? A search through the relevant literature (Bluestone and Harrison 1980; Erickson 1980; Healey 1982; Henderson 1979; Leigh and North 1978; Massey and Meegan 1978, 1982; O'Farrell 1976; Townsend 1983a) reveals that there are four (at least) different but closely related groups of influences which could be associated with such decisions. These can be labelled loosely as follows:

(1) *commercial* – influences related to specific market trends;
(2) *technological* – influences related to the productive capabilities of different plants;
(3) *organizational* – influences related to the historical and sociological ties which exist between different parts of corporations;
(4) *human relations* – influences related to the relationships between management, unions and local, regional or national government bodies.

The following sections of this chapter draw on a wide range of examples of closures announced since 1976 in various corporations, in an attempt to illustrate the possible origins of such job losses.

COMMERCIAL INFLUENCES

These operate as the 'umbrella' factors influencing the overall strategy upon which many closure decisions are based. Put simply, there are often strong reasons for closing capacity in weak markets while retaining plants operating in more lucrative (or stable) ones. The way in which this factor works is, however, quite complex: there are at least two other points to consider.

First, there are differences between the way in which activities are structured within different corporations. Erickson (1980) for example, investigated three such structures. *Conglomerate* corporations, with highly diverse and often unrelated interests in different markets, operate plants with relatively few exchanges between them. In such circumstances, it is possible to close some plants without affecting the operations of the remainder. *Single product* corporations, however, may have little flexibility in selecting plants on the basis of market trends, although there may be scope to withdraw from certain specific product lines to concentrate on more specialized areas of the same market. *Vertically integrated* corporations, on the other hand, display many inter-plant linkages, built up in past years through backward and forward integration as the corporation has attempted to gain control over the chain of production upon which it depends (Moore 1973). Closures in one part of the chain may therefore be restrained because they would disturb the overall balance of activities within such corporations, whereas others may instead focus on parts of the production chain where facilities are duplicated. 'Duplication' is quoted in many closure decisions.

The second complication arises out of interaction between corporations either in a competitive or collusive manner. Clearly, if one company closes capacity in one market, it leaves room for another company to survive and improve its profitability. Although this type of behaviour has been investigated in growth conditions (Watts 1980, 129–36), much less is known about the ways in which corporate 'survival' strategies in recession involve such interaction. The break up of Spillers' large network of bakeries in 1978 is a case in point. The prospect of alarmingly high losses led the company to draw up a deal with its two major competitors in the bread manufacturing market, Rank-Hovis-McDougall (RHM) and Associated British Foods (ABF) (*Financial Times*, 8 April 1978). The discussions resulted in the closure of 23 bakeries, the purchase of 7 others by RHM and 6 by ABF. Exactly how such interaction affected closures (or the likelihood of closures) in other parts of RHM and ABF is not clear. With regard to the Spillers closures, however, it seems quite likely that their selection was influenced to some extent by the existence of duplicate facilities; that is, in a market-oriented industry, duplicate facilities in a similar location within the surviving corporations.

There are other examples where the amount of collusion between producers in determining closures is more apparent and involves ongoing commitments. The British Steel Corporation (BSC), for example, were engaged in 1981 in a series of talks (Phoenix 2) with private steel producers to reduce excess capacity in the industry by agreement, rather than through a threatened price war. As a result of these talks, several private plants including Hadfields (Sheffield), Duport (Llanelli) and Round Oak, a joint BSC/Tube Investments company (Staffs) were closed (*Financial Times*, 23 April 1981). More recently, BSC and Johnson-Firth Brown reached agreement on the rationalization of steel forging capacity involving a threat to 3000 jobs (*Financial Times*, 21 August 1982). The precise nature of these types of agreements is obviously crucial to our understanding of the incidence of job losses and closures. The agreements, for example, could involve one corporation withdrawing from one market while their competitors withdrew from others. This will be particularly important in industries where economies of scale in production need to be preserved. This was an important consideration for BP and ICI, who recently agreed to streamline their UK plastics operations. In effect, the reorganization, which involved much exchange of individual plants between the two corporations, resulted in BP pulling out of the PVC business while ICI abandoned polyethylene production in the UK (*Financial Times*, 18 June 1982).

These discussions demonstrate clearly that the ways in which market forces influence the closure decision need to be analysed in the context of (i) the structure of activities within corporations and (ii) the various forms of interaction which could operate between corporations.

TECHNOLOGICAL INFLUENCES

The second group of influences on the closure decision is related to the technology of different plants. Again in simple terms, some plants may be closed because they operate equipment which is either out-of-date or in some other sense inappropriate to the needs of the corporation. Such differences usually have consequences for the relative productivity and profitability of plants.

The first point to stress is that 'profitability' of itself is no guarantee of survival. Plants may be closed when making profits (Bluestone and Harrison 1980), or kept open when making sustained losses, for the sake of some longer term goals in corporate strategy. Secondly, lack of 'profitability' is a fairly inadequate explanation of closures, for corporations have the ability (to varying degrees) to predetermine levels of profit either by redirecting orders to other sites or, more importantly, by discriminating against certain sites in allocating new investment (Bluestone and Harrison 1980, 16). This may apply in particular to the decline of manufacturing in older parts of cities. This is arguably a consequence of the process of 'de-skilling', withdrawing from sites with high labour skill inputs, and transferring production to newer plants where technological changes reduce the levels of skill required (Massey and Meegan 1979; Massey 1979). Successive rounds of investment may therefore have imposed low levels of productivity on certain plants (with consequences for profitability). Even if such discrimination has not occurred – and the present recession does *not* appear to concentrate closures on inner cities – it will often be the case that some plants may be due for re-investment, while others have fairly recently been modernized, the former being selected for closure. Courtaulds, for example, announced the closure of seven mills in the North-West involving the loss of 1200 jobs (*Financial Times*, 30 August 1980). The company had been modernizing parts of the Northern Spinning Division in the previous few years and 'was now concentrating its production in its newer units'. A similar consideration was involved in Tootal's closure of the Sunnyside Spinning Mill in Bolton (*Financial Times*, 5 April 1978), production being concentrated in five other mills which were 'much younger'.

Older plants have certain other characteristics which can reinforce their susceptibility to closure. In some such plants, buildings may be poorly designed for the needs of modern industry. Heinz, for example, closed their older plant near Wigan (*Financial Times*, 15 January 1979) according to management because its 'buildings [were] now considered unsuitable', transferring production to a more modern factory near the town at Kitt Green, which was opened in 1959. Many other older plants are naturally located in the conurbations, where room for expansion, loading and unloading is restricted. Furthermore, it is possible to offset part of the costs of closure when selling high valued city centre sites.

Modernity, however, is certainly no guarantee of survival, as some closures in the study period have involved the most recent facilities of corporations. Courtaulds, for example, closed its acrylic spinning mill at Spennymoor in County Durham (*Financial Times*, 20 January 1979) with the loss of 1560 jobs. The factory had been set up in 1969 in the Development Area, was extended in 1970 and again in 1973. Though not 'new' in a strict sense, the plant was certainly more modern than many other plants operated by the company. Courtaulds blamed the closure primarily on changes in fashions and falling demands for acrylic yarns, although labour relations had been poor: the closure of the plant had been preceded by a failure of management and unions to agree to changes in working conditions and redundancies. There is in general some suggestion that the most recent plants may be prone to closure possibly because they still have teething problems with technology, working practices and labour relations (Townsend 1983a, 68).

The technological capabilities of different plants may also be affected by plant *size*. Healey (1982, 40) suggests that small plants are more likely to close, partly because less capital and labour are tied up in them. The closure of several small plants also has commercial and public relations advantages in avoiding the publicity generally attached to large redundancies. In some industries small plants may be more likely to close because of the importance of economies of scale in the industry (Massey and Meegan 1982, 147). Empirical evidence tends to support these ideas. Healey's (1982) analysis of multi-plant enterprises in the textiles and clothing industry (with headquarters in the East Midlands, North-West or Yorkshire and Humberside) confirmed that closures disproportionately affected smaller plants. Henderson's (1979) results in Scotland reinforce this.

Large plants on the other hand may be more vulnerable to *part-closures*. Henderson's results show that while smaller plants have, as would logically be expected, a higher closure rate, larger ones are not less likely to shed workers, possibly indicating the number of part-closures which occurred. In practice, however, the distinction between a general slimming of a workforce and part-closures as discussed earlier is fairly fine and should be applied with caution.

ORGANIZATIONAL INFLUENCES

The impact of organizational differences between plants is even more difficult to analyse. Several studies have previously suggested that distance from a plant to the organization's head office has some bearing on closure, with more remote plants being closed rather than those in the head office region (Healey 1981). This argument is related to the way in which many corporations are presumed to operate branch plants, many of them in peripheral 'assisted areas' (Townroe 1975). As Watts (1981) has demonstrated, the evidence of such branch plant behaviour is extremely inconsistent. In some cases remoter branch plants are the first to be closed, but in others they represent very stable sources of employment. This inconsistency may be partly related to differences between the autonomy of branch plants, some being more closely integrated into the activities of the corporation than others. Erickson (1980), for example, has observed much lower closure rates for subsidiary plants operated by 'conglomerate' organizations compared with more typical branch plants which are strongly linked with other parts of the corporation.

Evidence of a more generalized nature however does suggest that peripheral regions of the UK are *first* to be affected by closures. Henderson established from the Records of Openings and Closures compiled by the Department of Industry that 'the closure rates are consistently higher than the national average in the peripheral regions' and that 'Scotland was higher than the peripheral areas as a whole' (1979, 13). Similarly, Townsend (1983a, 81) charted the announcement of closures in the twenty leading UK corporations using published reports in the *Financial Times*. Townsend observed that '13 of 16 groups *announced a loss of jobs in "assisted areas" before any in the rest of the country*'; in 1976–81 two further groups had job losses only in these areas and two announced their first losses in both types of area simultaneously. The reasons for this pattern are, however, not known in detail and the effects of 'distance from head office' remain speculative.

Other studies have suggested that closures are more likely to occur after a take-over of one company by another (Leigh and North 1978; Smith 1979). Here again, the early evidence is not consistent with popular belief that acquisition leads to mass closures. Leigh and North, for example, noted that acquisition 'more often resulted in expansion of output from acquired plants' (1978, 173). Assessment is not helped one way or the other by the fact that closures announced in the *Financial Times* between 1976 and 1981 barely mention this practice. The activities of the Bernard Wardle Group are exceptions to this generalization. Some controversy surrounded their closure of a PVC factory in Caernarfon, upon which production was to be transferred to its Armoride factory at Earby, Lancashire, which had been acquired only two years previously (*Financial Times*, 11 February 1980). The closure was announced just before the Group itself was acquired by Birmingham and Midland Trust (*Financial Times*, 28 March 1980). Later, the Bernard Wardle Group purchased Storeys Industrial Products (SIP) in 1982 and subsequently closed SIP's factory at White Cross, Lancaster, with the loss of 650 jobs (*Financial Times*, 18 December 1982).

THE INFLUENCE OF HUMAN RELATIONS

Finally, closures, particularly those involving large job losses, tend to be influenced by the pressures of industrial relations and a country's 'territorial politics'. Confrontations between unions and management can work two ways. Some corporations may close plants where labour relations are poor as, for example, when BOC closed its Transhield depot on Merseyside after an unresolved ten-week industrial dispute (*Financial Times*, 31 May 1979). Similar circumstances surrounded BL's closure of its car assembly plant at Speke, also on Merseyside. On other occasions, however, plants where little opposition is expected may be selected for closure, although examples where this applies are naturally difficult to find. Some corporations may also be tempted to retain capacity in places where high unemployment rates enable them to introduce changes in working practices without opposition (Massey and Meegan 1982, 149). Relations between management and government can be of considerable importance, particularly in areas of high unemployment, though recent evidence is limited. Governments have in the past gone to considerable lengths to delay or alter closure decisions, as we shall see in the last section of this chapter.

This discussion has served to illustrate the complexities associated with just one type of job loss, that related to closures of plants. Table 3.2 attempts to summarize the various arguments developed in the last four sections, listing some of the main dimensions on which closure decisions can be analysed. The ideas presented, though based on some empirical evidence, are couched largely in theoretical terms. Forthcoming publications, based on case studies of three

Table 3.2 A summary of the dimensions for the study of closures

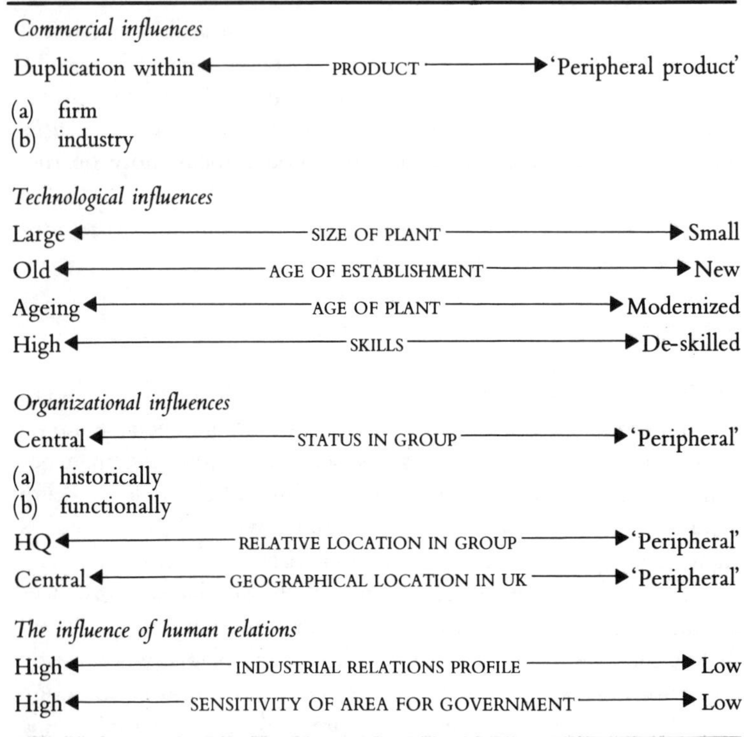

major corporations in the metal-using sector of the economy, will demonstrate how these groups of influences on closures can be separated in reality.

Policy issues

The methodology outlined in this paper has already suggested certain priorities for research in relation to policy. Our approach does not merely search for an understanding of the influences affecting corporate decisions related to closures, but, by doing so, provides a framework within which seriously to *question* the validity of such decisions when weighed against the costs to local communities. In this sense, our policy suggestions are related to specific events affecting particular plants but viewed from the perspective of the whole corporation. The very existence and structure of our policy suggestions themselves arise from the availability to us of data of unusual structure (plant rationalization which we have structured in terms of controlling financial groups). Research should focus more on the analysis of changes in *named* corporations, selected on the basis of their importance in terms of employment, their dominance of particular sectors, or their past and current behaviour (not least, those where closures of plants have already been undertaken, which are often part of an ongoing sequence of decisions). This final section builds further on these suggestions and considers specific areas of research which have particular relevance to policy. Before exploring new policy measures, it is useful to recall the Conservative government's attitude to workforce reductions, and the existing planning structures within which future policies are likely to be channelled.

WORKFORCE REDUCTIONS: THE EXISTING STRUCTURES

The questions of 'lame ducks', of subsidies to industry, of nationalization and privatization, and of selective technological development must remain at the centre of UK thinking for the next decade. Most of these issues are of great interest to unions as well as management and government, and this will emphasize their importance further in any 'tri-partite future'. *This is all to say that workforce reductions are an intensely political subject, and in some views they are the principal tool used by the Conservative government of 1979 – in changing the overall relations of capital and labour*. At the level of casework concerning actual proposals for particular plants, we may remind ourselves of the range of central government interests.

(1) *Cabinet level*: intervention at this level generally involves cases where many jobs are at stake; for example, all reports at end-1982 agreed that BSC wished to close one of their five GB integrated steelworks, and that it was a Cabinet decision to rescind this proposal. In potential political significance this

was not unlike the 'crisis' of Upper Clyde Shipbuilders in the early 1970s, then privately owned.

(2) *Departments responsible for nationalized industries*, including BSC or British Shipbuilders, are involved with rationalization of facilities through a variety of controls.

(3) *Department of Industry* has an interest in most types of industrial change and might be approached for – or want to intervene with – assistance via the Industry Act, 1972, whether or not this was through the clauses covering regional policy; it is of course arguable that any significant redundancy in a multi-plant firm is of interest for regional policy.

(4) *Department of Employment* requires prior notification of all redundancies occurring in batches of 10 or more (these require 30 days' notice; 90 days are required for batches of 100), and are responsible for the state contribution to redundancy payments (Townsend 1983b).

(5) *Manpower Services Commission* is responsible for a wide range of employment services which are relevant to redundant workers, for most types of 'job creation schemes', and for the Temporary Short-Time Working Compensation Scheme.

(6) *Local authorities* are increasingly involved in remedial action, chiefly with small and medium-size plants.

There is now an international body of government experience concerning *Workforce Reductions in Undertakings* (Yemin 1982). It is not the concern of this chapter to deal with general policy under headings (4) and (5) towards the general mass of employment change across the whole economy. Half a million redundancies have recently been notified per year, many of course in small batches and in single-plant enterprises, which are beyond the scope of this chapter. Our aim here, rather, is to concentrate, as in the rest of this contribution, on major redundancies and/or those in multi-plant corporations, bearing in mind these enterprises' past role in (partially) fulfilling the aims of regional policy.

THE MULTI-PLANT CORPORATION AND THE CASE FOR AN ACTIVE POLICY

Despite the apparent wide-ranging nature of the various levels of intervention described above, spatial policies with regard to closures in particular, and redundancies in general, have largely been *remedial* in nature; they have been designed to delay and/or reduce the levels of job losses in the fairly short space of time available between the announcement and implementation of redundancies. This section argues for a more constructive and *active* role for government in relation to job losses, geared more to *preventing* those circumstances in which particular labour markets suffer an irretrievable *net* loss of jobs. The assumption made here is that a 'regional policy' is still desirable, despite the implications of high unemployment and low investment throughout the economy (Townsend 1983a, 209–13). Like Massey and Meegan in chapter 5 we believe that a national decline of manufacturing employment is not a necessary cause of the decline of regional policy; such a policy is still technically possible, although working within restricted and modified circumstances. The main problem is political (Massey and Meegan 1982, 201).

The need and scope for government intervention may be at their greatest in multi-plant corporations. If we imagine a corporation which for capacity reasons has to close one of its five major plants, there may be a variety of possible scenarios. These might range from the situation where the corporation's internal analysis, *and* an analysis of social costs, pointed to the closure of the same plant. There might, however, be the situation where direct cost forecasts pointed to one plant, and calculation of social cost to another. There might finally be the situation where calculation of commercial and social costs provided no clear answer as between two plants – and the decision was purely one of territorial politics, involving conflict between the workers of different areas.

For the vulnerable area we are now talking of preventive as opposed to remedial policies. The redundancy payments legislation operates principally in terms of remedial action. Prior notice of 90 days for larger redundancies is designed to encourage the employer to find jobs for redundant workers elsewhere in the same organization. Equally, the notice may serve as a prerequisite for the provision of government compensation for temporary short-term working, or more generally for the deployment of MSC's employment services, often through visits to the plant. *It is proposed here that the concept of advance warning is extended to say six months for larger redundancies, and consideration given to government controlling the event, in larger corporations with greater resources, through being able to refuse permission for closure*, until at least appropriate conditions are met.

This proposal is not intended simply to allow more time to transfer or absorb job losses elsewhere, but to enable serious consideration of alternative courses of action to avert the threatened closure. Ninety days, for example, is far too short a period to permit preparation of reports on the social cost of closure or the potential for alternative job losses at other sites. 'Social audits' of this type have been prepared in connection with some previous closures. Wear Valley District Council and Durham County Council (1983), for example, prepared an analysis of the possible costs of closing Wolsingham Steelworks, announced by British Shipbuilders on 5 January 1983. Such audits include consideration of keeping open an otherwise unprofitable plant (through certain state subsidies) on the grounds of high social costs in terms of redundancy payments, unemployment benefits, loss of revenue from taxes and community losses causes by deflated local demand. Inquiries could also be made into alternative redundancies at other plants within the same corporation where such costs (and overall net costs) might be appreciably lower. This proposal too has precedent in the British mixed economy. It seems clear for instance that the pattern of closures in the 1960s rationalization of British Railways' workshops protected those in areas of higher unemployment (for example, Shildon) rather than other sites which closed. It is very likely that Department of Industry 'package deals' with corporations in the negotiation of industrial development certificates involved conditions over the retention as well as opening of plants in assisted areas.

Longer warning time would also encourage more serious consideration of 'buy-outs' by existing managers and/or workers. Several examples can be quoted of successful buy-outs of factories by former managers, securing at least some of the previous employment. Brigham and Cowan Shiprepair yards, Hull, closed by British Shipbuilders in 1981, reopened in July that year under the ownership of three of its former management (although with much lower manning levels) (*Financial Times*, 7 July 1981). BL accepted an offer from management at the Rearsby Components factory in Leicester in 1981 (*Financial Times*, 13 October 1981), while Stone-Platt sold their electrical division to its former management (*Financial Times*, 28 May 1982). Such experiences also have precedents in the US where 10,000 employees at the Weirton West Virginia plant have been negotiating to buy the factory from the National Steel Corporation.

Employee purchase adds a new dimension to such practices, suggesting that the future of some factories can be secured by more employee involvement in management decision-making. Problems associated with management and market forces, however, may ultimately militate against the survival of even such co-operative behaviour. Mr Anthony Benn set up several co-operatives of this nature in 1974 (as Industry Secretary) – Scottish Daily News, Meriden Motorcycles and Kirkby Manufacturing and Engineering (KME) on Merseyside. By 1978 two of these had closed and a third was in financial difficulties, indicating how vital it is to establish national and international conditions conducive to employment as well as working from local levels. Policies of this nature rely heavily on liaison between governments and industry.

A third option might involve a consideration of non-profit production for a plant under threat of closure. As one example among many, such plants could be modified to produce specialized equipment for the Health Service where technically possible. To some, this proposal may seem unworkable, until one considers the 'socially beneficial' subsidies which are given to British agriculture as a means of stabilizing the supply of basic foodstuffs.

How then can such proposals be introduced? Before the present recession both Watts (1980, 278–9) and Cross (1981, 129) saw that corporations' closure proposals could be so arbitrary as to require state control. The legislative route could have several points of reference. Bluestone and Harrison (1980), for example, suggest a *year's* warning of closure instead of the UK's present (90 days) prior notification for redundancies numbering more than 100. They also suggest a ban on production transfers from one site to another, and compulsory repayment of funding and tax abatements received.

To impose legislation on *any* closures, however, seems unwieldy, and there may be grounds for suggesting that penalties and restriction should be stiffer for *larger* redundancies, taking into account also the alternative employment prospects in the same labour market area. In this way a *longer* advance notice would be required, for example, in the case of BL's Speke closure (3000 redundancies) compared with BL's redundancies at their other Speke factory (500) announced in 1978, or their closure in Southall, London (2500 redundancies).

Any suggestion of legislation of this type has, in the past, not been well received. The Trades Union Congress, when consulted over redundancy legislation in 1965 and 1977, failed to persuade two successive Labour governments of the need to control redundancies (Yemin 1982). It seems from the whole of US state experience that there is 'little likelihood that the proposals for legislation intended to restrict plant closing will make much headway, especially today when law makers give inflation fighting a higher priority' (Yemin 1982, 190). Legislation of this type is however problematic in that no government can unilaterally restrict *multinational* activities without inviting boycotts from multinational corporations' finance and investment. Even controls placed on UK-based companies may have the wrong effect, encouraging them in the absence of international exchange controls (abolished in 1979) to invest ever more heavily overseas. Furthermore, even with such legislation in existence, there would be no compulsion for governments to enforce it rigorously.

The measures described above are all arrangements which would *post-date* the employer's decision to reduce capacity. It remains to stress that preparatory planning may be possible before even the firm's decision is made. This should be undertaken through government's forward action in contact with employers, and through research – the latter in producing a forecasting capacity in this field through use of a data-bank of existing closures. It is hoped that the present project of the authors (see acknowledgements below) will provide the elements of one such data bank to be available in the ESRC Survey Archives. It is expected that the analysis of individual corporations and 'bundles' of corporations will enhance our knowledge and understanding of corporations' selection of sites for closure on the kind of dimensions developed in pp. 73–80: for example, size of plant, age of establishment, or relative location within the corporation's group of plants. It is recognized, however (Massey and Meegan, chapter 5 of this volume), that existing records will *show a wider behavioural variation of response to economic pressures, that the nature of job loss in a given industry will change, and indeed that this is not a good field for the making of deterministic laws.*

It remains for government seriously to consider restoring physical controls over the location of industrial building, to consider legislation over the closure of plants, and to improve its knowledge of major corporations through requiring discussion of the role and future of all sites in multi-plant corporations. It is surely within the established sphere of British experience and politics to require forward confidential notification of major redundancies, sufficient to allow preparation of sites for replacement industries and the negotiation of proposals for incoming firms, before redundancy is even announced.

Acknowledgements

This work was partly funded by the Economic and Social Research Council, 1982–4, on 'The corporate sector and employment decline in UK sub-regions, 1976–81'.

4
Editorial introduction

The emphasis and approach of this chapter are very different from those of both the preceding chapters. Fothergill and Gudgin do not reject the analysis of individual firms or major corporations, but the role, form and purpose of such analysis is understood by them in a different way. It is through extensive, and comprehensive, analysis that Fothergill and Gudgin seek to identify the causes of urban and regional differences in employment change. It is through tenacious and step-by-step analysis of employment changes, disaggregated by type of area and industry, that the authors aim to track down the causes of geographical disparities. There are a number of distinctive characteristics which differentiate this approach from those in other chapters.

The approach aims to be comprehensive, and one implication of this, for the authors, is that the data themselves should be as full as possible in their coverage. Partial or sample data are less satisfactory, the reason for this being that there may be enormous variation both between industries and between small areas, and any conclusions drawn may therefore not be generalizable. Case studies present the extreme example of such problems of partiality: 'case studies should normally be avoided because it is difficult to know the extent to which the cases are typical, and the variability between cases often obscures trends that are clear when dealing with aggregate figures'.

As we saw in the discussion of chapter 3, the 'problems of sampling and of typicality' arise from, or in the context of, a particular conceptualization of the overall issue. Fothergill and Gudgin intend some of their conclusions to be empirically generalizable, and also, in their broad framework, to remain true over time. This is reflected in their clear and precise statement of aim. It is neither time- nor place-specific (save perhaps that it is bounded in its application to other advanced capitalist economies).

Following directly on from this, they argue that it is possible to set up testable hypotheses about causes, the evidence from which will be evaluated in terms of observable commonality or frequency of occurrence. It is important to be clear about terminology. Here a hypothesis (or a theory) is an explanation (or an attempt at one) through the identification of factors which can be observed as accounting for the outcomes apparent in the real world.

Fothergill and Gudgin emphasize the complexity of the situations they are analysing and point to the likely coexistence of a variety of sub-processes. In order to enable them to capture the complexity and yet to remain comprehensive they adopt a system of accounting as an overall framework in which to order their data. The question this raises is how the data should be disaggregated. They argue that the categories should, where the state of existing knowledge allows, reflect the proposed explanation but that 'if the theory is very tentative, or if several theories need examining, the data should allow disaggregation along several different lines'. The crucial and interesting question here is whether or not this disaggregation of data will come up with categories which relate conceptually to the underlying processes unearthed, for instance, by Lloyd and Shutt. To what extent is it possible, in other words, for the explanatory factors identified through this kind of extensive analysis to be conceptually coherent with the causes highlighted by other approaches? Fothergill and Gudgin are clear that their approach does not seek 'insight into how [particular] firms alter their production and employment in response to changing economic circumstances'. For them, the firm-oriented approach only comes into its own when aggregate numbers are the potentially unrepresentative product of a small number of cases 'when the law of averages ceases to be relevant and specific events no longer cancel out'.

The question which Fothergill and Gudgin address is explicitly spatial; they separate the issue of 'why are jobs being lost in this industry?' from the spatial one of 'why are more of the job losses occurring here rather than there?' And it is the latter alone on which they focus. They argue that trends in the national economy set the context for regional growth and decline, and affect areas differently, but they set out to analyse the spatial consequences, not the national trends themselves. Further, in framing their policy recommendations they make the assumption that the capitalist economy is likely to be with us for some time to come (not something with which the other contributors to this collection would disagree). For Fothergill and Gudgin, the conceptual implication is to 'take the existence of the capitalist system as given'. These two characteristics together mean that Fothergill and Gudgin pay less attention in their analysis to the internal organization of production on the one hand and to the form of social relations on the other than do the rest of the papers in this collection.

Putting all this together means that the explanation offered by this contribution is quite different from that in other chapters. It argues that changes in the location of manufacturing jobs in post-war Britain can be explained by a series of factors – urban structure, industrial structure, size structure and government regional policy. These factors interact with national trends to produce specific outcomes which may change over time. But the factors themselves are seen as being empirically generalizable, from place to place (including from country to country) and from one historical period to another. Finally, and again deriving directly from their method, these factors refer, not only to characteristics of industry or corporations, as in the other contributions, but to characteristics of *regions* which are relevant to industries.

Unlike the other chapters, this contribution argues that there is no direct link between politics and method. Fothergill and Gudgin suggest that politics often determine the issues that are chosen for investigation, but that recommendations do not flow automatically from research findings, whatever the method that has been used to arrive at those findings. In their view, recommendations depend a great deal on wider policy goals and political perspective, so that both right-wing and left-wing recommendations can sometimes be drawn from the same findings. In framing their own recommendations, Fothergill and Gudgin argue for a *regional* policy which must change the way it operates to arrive at spatially and socially more acceptable outcomes.

4

STEPHEN FOTHERGILL &
GRAHAM GUDGIN

Ideology and methods in industrial location research

Introduction

Disparities in employment opportunities between peripheral regions and southern England, and between inner cities and other areas, are enduring characteristics of the British economy and are likely to remain prominent features for at least the remainder of this century. These persistent imbalances in regional and local labour markets are justifiably an important political issue. However, there is no agreement about the nature and causes of the problems. Are multinational companies to blame or are the shifts taking place the inevitable result of changes in technology? Are depressed areas so hard-hit because they are 'unsuitable' for modern industry or because of the particular nature of Britain's industrial decline? Failure to resolve these and other questions has made it difficult to develop appropriate and effective policies. Too often, policies have provided short-term palliatives rather than lasting solutions.

Urban and regional policies are likely to be more successful if they are based on a sound understanding of the causes of disparities in growth. The main purpose of this chapter is to set out an approach to achieving such an analysis. The methodological guidelines that are put forward are derived from the authors' own experience in undertaking research on the location of employment change. Nevertheless, it is our contention that the approach we advocate is widely applicable and produces results that are incisive, robust and a basis for further research.

The first part of the chapter briefly outlines the main questions our research has tackled and the contribution we think it has made to an understanding of urban and regional growth. This research is chiefly reported in *Unequal Growth* (Fothergill and Gudgin 1982) and in several earlier and more recent publications, notably Gudgin (1978) and Fothergill, Kitson and Monk (1985). The second part describes the aspects of the methodology that have been critical in producing fruitful results and then illustrates our 'guidelines' by discussing their application to two different research questions. This is followed by comments on one of the main alternative approaches to industrial location research. Finally, the chapter considers the role of ideology in urban and regional research and the link between conclusions and policy recommendations.

An outline of our research

An important point to make straightaway is that though our research has been funded from a number of sources and undertaken at several institutions over a decade or more, the deliberate aim throughout has been to identify the causes of urban and regional differences in employment change. At first sight this may appear a grandiose aim, or one which is too general to be meaningful, but it is necessary to go back only as far as the 1960s to understand why such an apparently ambitious goal was required.

Present knowledge about urban and regional development may be inadequate, but ten or more years ago the situation was much worse. The location theories of Losch and Weber, with their reliance on transport costs, still dominated industrial location theory but were grossly out-of-step with the real world of the late twentieth century. The multinational company, the motorway network, the persistence of regional imbalance and the decline of the industrial city had little place in this theoretical world. On the other hand, empirical research had not provided a satisfactory alternative. Empirical work was going on, but a great deal of it by geographers was descriptive and economists largely shunned regional work. Furthermore, much empirical work was 'local' in its focus, so it was difficult to draw general conclusions. The big issues – why there are large and persistent regional differences in growth for example – were never tackled satisfactorily, and when these issues were addressed, as in *Regional Policy in Britain* (McCrone 1969) and *The Framework of Regional Economics in the UK* (Brown 1972), progress was strictly limited by an inadequate empirical base.

Our starting point was to reject most existing industrial location theory as patently irrelevant. *In its place we have developed an alternative view of urban and regional growth which not only rests on empirical evidence but is also comprehensive, in that it attempts to provide a framework for interpreting trends in all sectors of the economy and all locations. It also tries to explain how and why the pattern of employment change responds to changes in the national economy.*

A fundamental aspect of our approach is the division of local economies into two sectors – a 'basic' sector which leads economic growth in an area, and a 'dependent' sector serving local markets, whose growth depends on the growth of the basic sector. By and large, primary and manufacturing activities are basic and services are dependent. In other words, most service jobs follow the location of primary and manufacturing jobs, and do not normally act as an independent motor of local economic development. This is particularly true of private services which depend on local markets, but in many public services such as health and education the scale of provision in any area depends on population levels and thus on the growth or decline of the rest of the local economy. There is nothing new in these ideas. Our contribution has merely been to demonstrate the continuing relationship between these sectors during the last three decades.

Explanations of urban and regional employment change therefore need to concentrate on basic activities and especially on manufacturing, which is much the largest of these. *We put forward a 'structural' explanation for the location of growth and decline in this sector. By this we mean that in any given national economic context the spatial pattern of employment change is dependent on the particular urban and industrial characteristics which different areas have inherited from the past*. Three structural characteristics are important.

The first is *industrial structure*, or the mix of industries in an area. In the country as a whole, different industries experience different rates of employment change. This reflects a number of factors – changes in demand, the growth of labour productivity, and import penetration for example – and results in spatial variations in employment change because the growing and declining industries are not evenly spread across all areas. An area dominated by nationally declining industries, for instance, tends to experience large job losses.

The second is *urban structure* – the extent to which an area is urbanized. During the last twenty years Britain's cities have experienced a massive loss of manufacturing jobs while small towns and rural areas have been successful (at least until the recession at the start of the 1980s) in retaining and expanding their manufacturing employment. As a general rule, the larger the settlement the faster the decline.

The third structural element is the *size structure* of factories. This influences employment because it is the main influence on the location of new firms. Despite their tiny initial size, new independent firms make a significant contribution to employment, partly through weight of numbers and partly because on average they experience healthy growth during their early years. A disproportionately large share of them are set up by people who previously worked in small firms, so areas with a substantial heritage of small firms experience higher rates of new firm formation than areas dominated by large factories.

These three structural characteristics are the product of many decades of investment and they change extremely slowly. Collectively, they exert a dominant influence on underlying trends in the location of manufacturing. *Regional policy* is the fourth major influence on urban and regional employment, but one which differs from the others because it is a response to slow growth rather than a cause of underlying disparities. Regional policy has led to a substantial shift of manufacturing jobs into the assisted areas.

The usefulness of this framework for understanding urban and regional employment change in manufacturing is illustrated by the contrast between the East and West Midlands. To many people, the division of the English Midlands into two regions represents an administrative convenience rather than an economic, social or physical reality, and to some extent this view is justified because in practice the Midlands comprise a series of city-regions rather than two cohesive, functional units. The apparent similarity between the two regions is enhanced by the fact that until the beginning of the 1980s both seemed relatively prosperous, with unemployment rates normally a little below the national average. Yet despite being adjacent regions with much in common, the contrast in employment change is considerable. For example, between 1960 and 1975 the West Midlands lost more than one in eight of its manufacturing jobs while manufacturing employment in the East Midlands remained more or less unchanged. Traditional theories of urban and regional growth are at a loss to explain how this contrast arises. The structural framework we have outlined provides the necessary insights. Let us take each factor in turn.

- Industrial structure favoured the West Midlands during the 1960–75 period. This was mainly because the motor industry (a major employer in the region) was still fairly buoyant during these years, whereas employment in the East Midlands was handicapped by the importance of declining industries such as footwear. On balance, we estimate that the West Midlands' industrial structure boosted the region's manufacturing employment by 6–7 per cent relative to its neighbour.
- Urban structure in contrast favoured the East Midlands. The West Midlands is one of Britain's most urban regions with a high proportion of its manufacturing in the Birmingham conurbation and two further concentrations in Coventry and Stoke-on-Trent. The East Midlands includes Nottingham, Derby and Leicester, each only a fraction of the size of the Birmingham conurbation, and a higher proportion of its manufacturing is in small towns and rural areas. As industry throughout Britain has declined in major cities and grown in small towns, the East Midlands has therefore benefited. We estimate that the balance of cities, towns and rural areas in the two regions was responsible for 12–13 per cent better growth in manufacturing employment in the East Midlands between 1960 and 1975.
- Size structure also favoured the East Midlands. This region

has a substantial heritage of small and medium-sized firms, whereas in the West Midlands a larger share of employment is concentrated in big motor and engineering plants, especially in Coventry. The more favourable size structure in the East Midlands, and the consequent higher rate of new firm formation, is estimated to have boosted its manufacturing employment by 4 per cent relative to its neighbour between 1960 and 1975.

- Regional policy led to small job losses in both regions. With one or two minor exceptions, the East and West Midlands have not received any regional aid, and both have consequently lost jobs, mostly by the diversion of growth into branch plants in the assisted areas.

The combined influence of these four factors explains why manufacturing employment in the East Midlands has fared so much better than in the West Midlands, despite the physical proximity of the two regions. The East Midlands' less favourable industrial structure was more than offset by the better growth associated with its urban structure and size structure. That it is possible to understand the difference between these two regions – or between other regions for that matter – without reference to traditional theories of industrial location, casts grave doubt on the validity of those theories as explanations for regional growth and decline.

It must be stressed that structural characteristics do not by themselves generate spatial variations in employment change. It is the interaction between changes in the national economy and the structural features of areas which produces divergences in local trends. Industrial structure, for example, is an important influence only because there are differences between industries in the national rate of employment change resulting from variations in the growth of demand, labour productivity and import penetration. Similarly, the urban structure of an area has been a determinant of growth only because of the national increase in manufacturing's floorspace requirements, which could not be easily accommodated within major urban areas. If manufacturing had not required large additional amounts of floorspace, the pressures generating an urban-rural contrast in employment change would have been much smaller.

The interaction between national trends and the structural characteristics of areas is important because national trends are not stable. The stimulus to the West Midlands provided by the motor industry has disappeared now that this industry is in serious national decline, and regional policy no longer provides much benefit to the assisted areas now that fewer firms are opening new factories that might be diverted there. These are two examples of the process through which changing national trends affect local areas. The structural characteristics of areas remain unaltered, but the way in which they mould the pattern of urban and regional development depends on national industrial trends.

Methodological guidelines

As will be apparent by now, our view of the causes of urban and regional employment change owes hardly anything to traditional theories but is instead rooted in empirical findings and an understanding of the organization and structure of industry. In coming to this view we have followed a number of methodological guidelines, some of which are only clear with hindsight, which we think are widely applicable in industrial location research.

(i) *The question must be defined clearly.*

This is an obvious point, but it is surprising how much research seems not to have a clear goal. Description rather than explanation is often the result in such circumstances. A common problem, for instance, is the failure to decide whether the central question is a national one (e.g. why are jobs being lost in this industry?) or a spatial one (why are more of the job losses occurring here rather than there?).

Most of our research has taken national trends as given and has not sought to explain them, but instead has concentrated on identifying their urban and regional consequences. Thus in looking at the effects of industrial structure we tried to assess how national variations in growth between industries affect different localities.

(ii) *The main characteristics of the problem should be investigated.*

At an early stage in most well-designed research a theory is developed to explain a phenomenon that has been observed. The theory is then tested by collecting the data necessary to examine one or more hypotheses derived from the theory. Generating meaningful hypotheses is difficult however. This task is greatly eased if the research includes a stage prior to the formulation of the theory: the assembly of a variety of information on the characteristics of the phenomenon to be explained.

In some research fields this stage may have been reached by previous work, leaving only a final synthesis to be achieved. However, like most fields which have not attained the disciplined and cumulative advancement of knowledge that characterizes physical science, urban and regional analysis always requires large amounts of gap-filling empirical investigation. The availability of a range of detailed information is indispensible when the time finally comes to formulate a theory. In particular, because a great deal more is known about the problem the theory must be formulated more precisely in order to be compatible with several aspects of the problem. It is therefore less likely that the theory will be substantially incorrect, and a great deal of wasted time spent testing an entirely inappropriate theory can normally be avoided.

Of course the act of gathering information on the characteristics of a problem cannot be undertaken in the absence of ideas about the processes that may be at work and the sorts of information that may be most revealing. But it would be wrong to elevate such 'ideas' to the status of 'theories'. They may be only hunches. There is no reason why information relating to several conflicting ideas cannot be assembled at this stage, and the possible relationships between such ideas need not be thought through in detail.

A related point is the need to examine trends at more than one spatial scale, because a contrast observed at one scale may reflect processes which actually operate at a quite different scale. Until the late 1970s two separate strands of research on employment change – one 'urban', the other 'regional' – proceeded largely independently, each attempting to explain spatial variations in terms of 'urban' and 'regional' factors respectively. What has been realized subsequently, in part because of our own findings, is that regional differences in growth are to a large extent reflections of much larger urban–rural differences. Predominantly urban regions tend to decline; rural regions grow.

(iii) *Existing theories should be questioned and new ones derived from evidence rather than untested assumptions.*
A surfeit of existing theories or approaches is usually available for the researcher to develop and adapt. But existing theories ought not to be treated with undue reverence if they are at odds with even only a few pieces of empirical evidence.

There is no simple method by which a number of separate facts can be reconciled within the framework of one theory. A major influence is the overall paradigm within which the research is undertaken. In Britain during the last twenty years, the dominant paradigm has been broadly Keynesian. This views a region's employment as growing or declining in response to the demand for its products and services, in the same way as the national economy responds to aggregate demand. The popularity as a research tool of 'shift-share analysis', for example, rests on the Keynesian view that changes in the structure of demand for products affect the location of employment because of variations in the mix of industries between areas. Neo-classical economics, the main alternative paradigm and one which has always been more popular in North America, stresses variations in costs and efficiency and the supposed adjustment of wages and prices in the face of alterations in supply and demand. More recently, Marxism has also been applied to urban and regional analysis, where the roles of class conflict and falling profitability have been given prominence.

We broadly accept the Keynesian explanation of changes in output and employment in the national economy, and this has been implicit in our research. However, in the regional context Keynesian economics provides only limited guidance, particularly since there are large changes in urban and regional employment which cannot be accounted for by changes in demand. These are measured for example by the large 'differential shifts' revealed by shift-share analysis. These differentials are the changes in employment which occur over and above the changes expected on the basis of an area's mix of industries and the national trends in employment in those industries. Hence there has been a need to advance beyond Keynesian ideas which, although vital to understanding employment change, provide only one link in a complex chain of causal influences. Neo-classical economics is usually little help. The sorts of adjustment of wages and prices which are supposed to occur simply do not happen to any significant extent in contemporary Britain.

Indeed, existing paradigms are too often accepted uncritically and used to guide research. Admittedly it is helpful to have an accepted body of theory to deal with difficult and confusing research questions, but unless the theory provides a reasonably accurate description of the world the gain is illusory. An eclectic approach, confronting a range of ideas with appropriate evidence, is better. New theories must be evolved primarily with a view to fitting the available evidence on the phenonemon to be explained.

(iv) *Where understanding is poorly developed, a comprehensive 'accounting' system is necessary.*

A complex phenomenon such as employment change is usually the outcome of several separate but interrelated processes. Therefore unless the information that is collected is comprehensive there is a danger that some processes will be overlooked or obscured. Comprehensiveness can be achieved by adopting an 'accounting' system in which individual components sum to the total, and this has the added advantage of providing a measure of the relative importance of each component.

For example, a piece of research might conclude that multinational companies are relocating jobs from one area to another. Even though this may be a new finding its importance is greatly enhanced if the shift can be related to the magnitude of the aggregate change that is occurring. Is the multinationals' contribution a small one or a large one? Equally, it is helpful to know whether the contribution is independent of that made by other firms. Do other firms show the same locational trends? Does the relocation of these jobs prevent or encourage other firms doing the same?

Comprehensive information systems are needed in order to relate the particular to the aggregate. In the analysis of employment change, this means an information system covering all sources of employment with as much detail as is potentially relevant and feasible, and where large amounts of information are involved an accounting framework is essential.

Simple accounting frameworks are commonplace; official statistics of employment by industry and region are an example. Forecasting models of economic activity are invariably based on accounting systems showing the contribution to changes in national income of individual components (investment, consumption, public spending, exports, etc.). Similar methods are necessary in explaining historic trends. An aggregate change in employment, for example, can in principle be disaggregated so that the causal factors relevant to each part combine to form an overall explanation. Unfortunately, in regional research the difficulty in obtaining suitable information all too often means that the considerable potential gains from comprehensiveness are lost. The difficulties can usually be overcome with sufficient time and effort, but the conventional organization of research into small teams and isolated individuals, each with few resources, is a formidable obstacle.

The skill in using an accounting system is in the selection of a meaningful disaggregation which reflects the processes of change that are occurring. If a theory has already been developed, this should guide the disaggregation. If the theory is very tentative, or if several theories need examining, the data should allow disaggregation along several different lines. This is one advantage of data on individual establishments, because the same information can be aggregated and classified by industry, size of firm, or corporate status.

If the level of demand for different products is thought to be an important factor in differentiating areas, the data are appropriately disaggregated by industry, since market constraints affect industries differently. If instead the problem is thought to be a resource constraint – a lack of land for expansion or a shortage of labour, for example – a different dissaggregation may be required. In the case of land constraints, for instance, it is useful to categorize factories according to the extent to which the site is built-up and built around. Where market or resource constraints do not differentiate areas, firms have greater discretion over the location of their production and an alternative framework is required. In this case, it is probably useful to classify firms according to aspects of their corporate status: new independent firms face different choices to existing firms, and small companies do not have all the options available to multinational companies, who can shift production between countries as well as between local areas.

(v) *The research question must br broken up into manageable parts.*
There is a fashion for econometric modelling which uses mathematical sophistication to try, in one set-piece analysis, to squeeze

more meaning out of data. If the analysis packages together a number of discrete processes, as sometimes happens, the results are exceedingly difficult to interpret. A more useful way to proceed is to break up the main research question into manageable parts. Wherever possible these should reflect distinct processes of change, and each can then be researched separately using the most appropriate data and methods.

The categorization of employment change, associated with a comprehensive accounting framework of the sort described above, in no sense provides explanations. However, *by adopting an accounting framework it is usually possible to break an overall question into several sub-questions*, and the framework provides a measure of the relative importance of each sub-question in terms of the number of jobs involved.

For example, the accounting framework known as 'components of change' disaggregates net change into the employment associated with the opening, closure, expansion and contraction of individual factories. Spatial contrasts in the growth or decline of existing factories are different from those of new factories, and from new independent firms in particular. It is therefore useful to separate new firms from the rest and investigate each group as appropriate. In this way a complicated problem is taken apart and the underlying causal influences can be untangled a few at a time.

The separation of net change into components only makes sense if the components are reasonably independent of each other. If the same factors affect all components or if there are strong interactions between them, separating firms into components may not help disentangle the influences. This is a problem in investigating the growth or decline of existing factories. The balance between employment change in expansions, contractions and closures in fact mainly reflects the size of factories: small factories are more prone to closure than large ones, but are more likely to expand if they survive. Areas with a high proportion of employment in small factories therefore experience high rates of closure and expansion but low rates of contraction. The magnitude of each of these components cannot therefore be understood in isolation from the others.

(vi) *The data must be appropriate to test the hypothesis.*

This final methodological guideline is crucial. Hypotheses cannot be tested properly unless the data are relevant to the question asked. If more than one hypothesis is to be tested, the data should be sufficiently flexible or wide ranging to enable several questions to be examined.

Data for the whole of a study area are normally preferable to data for a part. This is because there is enormous variation at the level of the firm, and data covering small areas, and hence few firms, are liable to reflect events in a few large factories rather than more systematic influences. For the same reasons, data for all industries are normally preferable to data for only one or two. *Case studies should normally be avoided because it is difficult to know the extent to which the cases are typical*, and the variability between cases often obscures trends that are clear when dealing with aggregate figures.

Application of the guidelines

What has been outlined is an empiricist's tool-kit of guidelines. Our view is essentially that there are few short-cuts to success. A great deal has to be known about the characteristics of the problem in hand. Equally, the decision making units (in our case companies) have to be understood – how they operate and what constrains their actions. It is vital to undertake research in an information-rich context. Only then can small advances in theory be made. Each step forward throws up new questions that may require months or years of additional investigation before enough new information is available to make a further small advance. This is surely the way that all science makes progress. Major advances, which from the outside appear to be tremendous leaps forward, are usually the last step in a complex edifice of research involving many people, perhaps spread across the world. Similarly, in social science fundamental treatises are generally the product of many years' study of how economies and societies work.

The guidelines described above offer a method for undertaking successful research on a wide range of issues. They need not be applied in the order they were listed, though obviously defining the question to be answered must always come first. In practice, research often follows a cycle: a question is defined, it is broken into its component parts, a theory is developed and data are assembled to test it, only to raise further unresolved questions which in turn must be dissected and analysed using new data. The value of the guidelines can be illustrated by describing two parts of our research.

(i) *The urban–rural shift*

The first is our investigation of the causes of the urban–rural contrast in manufacturing employment change. This contrast underpins the importance of 'urban structure' as a determinant of regional growth and decline. The problem was to explain the existence and magnitude of the large and persistent differences in the rate of manufacturing employment change between cities, towns and rural areas. We took the national decline in manufacturing employment as given, and looked specifically at why cities have been hit worse than other areas.

Since the urban–rural contrast had not previously been documented the next step, following the guidelines, was to gather further information in answer to several rudimentary questions. Could the contrast be observed in all the UK regions? Was it occurring in other countries? Were all industries affected? What other relevant economic indicators showed consistent 'urban–rural' differences? At this stage the adoption of an accounting system proved useful. For the whole country, official data allowed the disaggregation of net change in each type of area into that due to factory movement and that due to other components. For one region, the East Midlands, net change was further disaggregated into openings, closures, expansions and contractions, again by type of area, and also by size of firm, ownership type, and several other characteristics. This was enormously time consuming, since the establishment data-bank for the East Midlands had to be compiled by ourselves. The results were presented in Fothergill and Gudgin (1982, chapter 5) and revealed that the urban–rural contrast is extremely pervasive in terms of the sorts of firms involved and locations affected. The contrast does not arise, for example, solely because of shifts in the location of production by multi-plant or multinational enterprises. The results also showed that the contrast mostly reflects the location of growth, in new and existing factories, rather than the location of job losses through closures and contractions. Some ideas were ruled out, notably the suggestion that an unfavourable mix of industries is at the root of the cities' industrial decline.

What was now required was a theory to fit all the available evidence. A wide range of ideas were confronted with the facts. Most were inconsistent with what had been observed. The theory we eventually evolved explains urban–rural differences in manufacturing employment change in terms of two national economic processes. The first is the rising capital intensity of production, which leads to a reduction in the number of workers on any given area of factory floorspace. This fall in 'employment density' has averaged 2–3 per cent a year during the last two decades and exerts a major downward pressure on employment in all areas. The second process is manufacturing industry's rising demand for factory floorspace. Since the mid-1960s this has increased by nearly 1 per cent a year, despite the fall in manufacturing employment. However, a disproportionately large share of additions to the stock of floorspace is located in small towns and rural areas because of the difficulty of physically accommodating large increases in floorspace in major built-up areas. The shortage of space for expansion in cities affects existing factories particularly acutely because most are hemmed-in by existing development, but sites for new factories in cities are also difficult to assemble and costly to acquire and develop.

In essence, our theory therefore argued that job losses occur in all areas because of the reduction in employment density associated with rising capital intensity, but most of the offsetting increases in employment in new factories and factory extensions are located in small towns and rural areas because of the greater availability of room for expansion. In a number of respects the theory fitted the evidence. In particular, it offered an explanation for the pervasiveness of urban–rural differences, affecting a wide range of firms irrespective of ownership, and for the key role played by the location of growth as opposed to decline. Several other peculiarities which had been observed made sense once this theory had been developed. The lack of any urban–rural contrast among the smallest firms, for instance, could be explained by their mobility, which enables them to hurdle the space constraints preventing growth in less mobile larger plants.

What we now had was a well-developed theory inferred from a variety of information, but the theory included assumptions (about changes in the utilization of floorspace and about constraints on new building) which required verification. The empirical testing was again undertaken within an accounting framework, though this time the structure of the framework reflected the new theory.

The first stage was to quantify the two processes. As the theory anticipated, job losses associated with falling employment density were found to be almost evenly spread across all types of area. Employment growth associated with changes in the stock of floorspace, on the other hand, is concentrated in small towns and rural areas. The next stage was therefore to look more closely at the processes making up the net change in the stock of floorspace. This revealed that nearly all the urban–rural difference in floorspace change is due to the location of new units and factory extensions. Demolitions and changes of use make little contribution to urban–rural differences. The research question was thus both refined and broken up into manageable component parts. New units and extensions needed to be the focus of attention, and the hypothesis about shortage of room for expansion in cities could be tested separately for each of these components using the most relevant data.

Taking factory extensions first, data on the sites and premises (especially the room for on-site expansion) of firms in cities, towns and rural areas were required. This was obtained by postal questionnaire and by site visits for almost 2500 factories in a range of urban and rural areas in one region. The results entirely confirmed the hypothesis: after allowing for differences in growth attributable to the sorts of sites and premises occupied by firms, the residual differences in growth associated with location were statistically insignificant. In other words, factories in cities experience poorer growth than those in rural areas because they are more likely to operate in old buildings on cramped sites, not because cities pose some other handicap to growth (Fothergill, Kitson and Monk 1985).

So far as new factory units are concerned, the data required to test the hypothesis concerned the availability of land for new factory building in urban and rural areas. Since such data are not normally collated, details of industrial land availability were obtained from all local authorities in Great Britain. The data showed that in relation to the size of the manufacturing sector in each type of area, small towns and rural areas have vastly more land designated and available for new industrial development than the conurbations (Fothergill, Kitson and Monk 1985).

(ii) *The location of new firms*

A second example illustrates many of the same methodological points. This concerns our research into the contribution and location of new independent manufacturing firms.

The initial identification of these firms as a subject of concern was in fact a by-product of the application of the principle of completeness in accounting for employment change. Until the 1970s, neither economic theory nor political ideology gave new firms a prominent role. However, in collating data on employment in all manufacturing establishments in one region it became apparent that large numbers of new firms continue to be established, and over a long period these make a significant contribution to employment. Nothing in existing literature on industrial location had led us to expect this. The almost accidental nature of these findings demonstrates the value of complete accounting frameworks.

Large contrasts in formation rates were evident, even between neighbouring towns, but at first insufficient data were available to be sure of the pattern. The strong suspicion was that areas dominated by large firms had low rates of company formation. Subsequent data for a larger area confirmed this tentative observation. Formation rates are four or five times higher in towns with almost no large firms than in towns where large factories, with 500 or more workers, provide three-quarters of the jobs. These differences in rates of new firm formation are the reason why 'size structure' is a determinant of the location of manufacturing employment change.

The next research task was to explain this spatial contrast in new firm formation. Previous research had already identified some of the main aspects of new firm formation (Gudgin 1978). Founders mostly start their firms near to where they already live. Many of them are former manual workers, starting up in trades in which they have previously worked. Initial premises are usually cheap old buildings, and personal contacts are important in securing early orders. Established economic theory offers little guidance concerning this sort of entrepreneurship, and no single new theory had a clear advantage in explaining what could be observed. Instead, hypotheses were developed on the basis of the facts available. The major possibilities were that formation rates varied because of differences in the relevant experience offered to potential entrepreneurs by existing local industry, or that some constraint (lack of suitable premises, labour or orders) prevented people starting firms in large-plant towns.

The data necessary to test these hypothesis was gleaned partly from the establishment data-bank we had assembled, which included a reasonably comprehensive list of all new firms in one region over a seven-year period. Mostly though, new information was necessary, and a highly specific interview survey was undertaken to explore the background of founders and the constraints they had faced. This method was less than ideal, but was successful because the research question – to explain the contrast between large-plant towns and other areas - was so tightly specified. The conclusion, reached simultaneously by Johnson and Cathcart (1979), was that a disproportionately large share of founders previously worked in small firms. Towns with a large population of small firms therefore have high rates of new firm formation and the number of small firms is thus maintained and increased over long periods. This process helps explain the relative vitality of cities like Leicester, which have a high proportion of their jobs in small and medium-sized firms and which continually renew their population of firms and to some extent their products. In contrast, large-plant towns, such as Coventry, are vulnerable to external influences on their few very large employers.

The research on new firms is another example of the importance of breaking down a research problem into meaningful parts. The causes of urban and regional employment change cannot be understood if tackled at an aggregate level. A number of quite different processes are involved, including the location of new firms and the urban–rural contrast in the growth of existing firms. The crucial step is to separate these processes, which are submerged by aggregate figures.

An alternative methodology: the 'firm-orientated' approach

It is impossible for this paper to compare our approach with all the alternatives that have been tried in the last few years, but it is probably helpful to comment on a currently popular methodology which is sometimes viewed as an alternative to the one we have put forward.

This is what we will call the 'firm-orientated' approach, exemplified by Massey and Meegan (1982). It starts by examining the economic pressures forcing individual firms to change and adapt their scale and methods of production. The pressures are mostly national and international in origin, and include switches in demand, changes in technology and competitive pressures. The response of individual companies is then analysed. What Massey and Meegan have done is to devise a framework for understanding the responses. Firms can intensify production without changing technology (by speeding up assembly lines for example). They can introduce technical changes in production. Or they can rationalize (i.e. cut back or close) existing capacity. Which of these responses is adopted depends on the context in which the changes are made, including the structure of the industry. Massey and Meegan argue that the pattern of urban and regional employment change is the outcome of many hundreds of economic pressures, each specific to their point in history, specific to individual firms and industries, and each pushing and pulling in different directions. An understanding of the causes of spatial disparities therefore requires an understanding of economic pressures on individual firms. 'Areas' themselves are not an important ingredient in any explanation.

Massey and Meegan's method appears at first sight to be an alternative approach to the issues tackled by our research. In fact, it is an attempt to do something quite different. We have tried to account for the large shifts in industrial employment during the post-war period and have not sought to explain events within specific companies. We do not disagree with the points Massey and Meegan make about individual firms based on their detailed case studies. However, *many pressures and responses that can be observed at the level of the firm tend to cancel one another out in a large area, and these pressures and responses can thus be safely disregarded in any study of overall urban and regional trends since they do not make a contribution to the net differences between areas.* This is the law of averages at work, simplifying the job of the researcher. Instead, *the focus needs to be on those forces which systematically discriminate between areas.*

There have been pervasive shifts in industrial location affecting many industries for prolonged periods, in years of both boom and slump, and affecting small locally owned firms as well as multinationals. Moreover, some of the trends, notably the urban–rural shift, are occurring in several other western industrial countries in macroeconomic circumstances often different from those in Britain. *Such powerful shifts in industrial location need explaining by more than merely the temporary co-incidence of hundreds of separate processes in individual industries.* It is necessary to explain who so many firms respond in similar ways. Our view, as noted, is that the structural characteristics of regions themselves and the industry within them are needed to account for the pervasiveness and persistence of these trends. National economic pressures must be included in the explanation, but it is the interaction between these pressures and the structural characteristics of areas which generates disparities in employment change.

Where our approach is inapplicable, and where the firm-orientated approach comes into its own, is when the law of averages ceases to be relevant and specific events no longer cancel out. In small areas dominated by a few large firms, the reaction of those firms to economic pressures is all-important. Our research has helped explain why such areas remain dependent on a few large employers, but it provides limited insight into how those firms alter their production and employment in response to changing economic circumstances. Massey and Meegan provide a helpful framework for investigating such questions.

Different firms do respond in different ways to the same pressures. This has been used to justify the firm-orientated approach and to criticize research methods like our own, which rely on the grouping of firms or places into categories. Such criticisms nevertheless miss the point. Although there is diversity of experience at the level of individual firms, the value of categories and groupings is that they highlight the *constraints* on aggregate trends. Individual firms' experience varies; some firms in a given industry grow while others decline. But in aggregate the employment in that industry may be constrained by the extent of the market, so that one firm's success in bucking market pressures may have its converse in the decline of another whose sales it displaces. The important point, in this context, is that the level of employment in the industry as a whole is best understood by reference to the size of the market.

Similarly, the urban–rural contrast in manufacturing employment change is more easily understood by reference to constraints – and to categories of firms and places where these constraints apply – than to the diversity of experience at the level of the firm. As mentioned earlier, there is evidence that the level of manufacturing employment in Britain's largest cities has been constrained by the availability of industrial land and premises, and that the constraint grows tighter as technical changes in production reduce the number of workers per unit of floorspace. Consequently, in the 1960s and much of the 1970s, when the growth of the economy meant that these constraints were particularly pressing, the closure of factories in cities resulting from individual firms' strategies probably had little net effect on cities' employment. What probably happened was that closures released land and premises for the growth of other firms, whose expansion locally might otherwise have been frustrated by the constraints on the supply of space.

In such circumstances the firm-orientated approach to industrial location research simply misses the main issues. It explains why an individual firm or small group of firms acted in the way it did, but invariably it fails to reveal the factors which constrain and determine the overall level of employment in any locality.

Ideology, methods and policy

Let us now confront the issue of ideology in industrial location research. How much should political and ideological views be allowed to influence the choice of methods and recommendations on policy?

Ideology is a major factor determining what is studied, in that researchers' political views influence the priority they attach to different issues. At present in Britain, many of the left regard employment and unemployment as priorities for research; the right maintain a fascination for inflation and monetary issues. But at another level, ideology plays a substantial if less obvious role in the choice of research topic. Cause and effect are organized in systems of causation in which the effects of one process become the cause of another. In studying any given system of causation – such as the process of employment change – researchers have to decide where to 'break into' the system. This decision often depends on ideology.

Let us take a simple example, far removed from regional research.

If a person were to press a light switch, what would most people say caused the light to come on? We would argue that the answer depends on the circumstances. With limited information the answer might be 'because the switch was pressed'. However, if an electrician had just rewired the building, the answer might be 'because the electrician had finished'. If the electrical generator had just been invented the answer might well concentrate on this fact, and so on. Actually, the light comes on because of a complex conjunction of events, and in focusing on one event the rest are taken as given.

In industrial location research, and other fields, a similar thing happens. In explaining disparities in employment change some researchers look at the nature of the capitalist economic system as a whole. Their view is that if capitalism were replaced by a different economic structure the same disparities in employment change would not arise. Hence the cause is capitalism itself. An alternative position is to take the existence of the capitalist system as given, and to investigate why it operates the way it does. Taking this second approach, explanations for disparities in employment change are likely to be concerned with the decisions of firms or the characteristics of places, rather than the existence of the economic system.

Our research is in the latter tradition. It is obvious that a different economic structure would give rise to different trends and problems, but the present structure seems likely to be with us for a while yet. Our interest is therefore directed at the way in which the system works, and towards identifying the points where intervention or reform might be possible and effective. This is, of course, an ideological position.

Once these broad ideological decisions have been taken the remaining decisions in research are mainly practical rather than ideological. The question of whether to concentrate on individual behaviour rather than on the behaviour of aggregates, or on actual decisions rather than the constraints on decision making, are all practical ones which depend on exactly what the research is trying to achieve. The choice of technique should follow from the choice of research question, in that the technique must be the most appropriate to answer the question. Ideology and method are thus linked only indirectly. Regression, shift-share analysis, components of change and case studies are all tools which can be employed for a wide variety of purposes, and their successful application is a technical not an ideological matter. Whether useful conclusions emerge from them depends on how rigorously they are used.

The other point in research where ideology plays a part is when policy recommendations are drawn from the conclusions. Policy decisions on social and economic matters are inevitably political. More importantly, they are not tightly constrained by the results of research. An example of a purely ideological link between research findings and policy recommendations is the work of David Birch on job generation in the US (Birch 1979). He found that two-thirds of net new jobs came from very small firms. His recommendations were that since small firms are so important in job generation they should be encouraged still further, and his views were enthusiastically received in Britain by the small firms lobby and by Conservative ministers. But on the basis of the evidence available to him he could equally well have concluded that larger firms were in difficulty, and more should be done to help them instead. His recommendations were a clear reflection of his own ideological position and were only tenuously related to his research findings.

Contrasting government reactions to the same urban and regional problems also illustrate the role of ideology. Labour governments have intervened in company location decisions, either directly through controls such as the Industrial Development Certificate or indirectly through grants and subsidies. Conservative governments have been less willing to take this course, and increasingly have eschewed intervention altogether. Instead, the Conservatives favour solutions through the operation of market forces, such as lower wages to stimulate competitiveness and employment. They also stress migration as a solution to unemployment, despite the less than obvious abundance of jobs in almost all areas. Solutions, it is argued, lie with individuals not the state. Because non-intervention is central to modern Conservative philosophy, for reasons mostly unconnected with urban and regional matters, consistency demands that urban and regional problems should be handled within this framework.

Our research has highlighted the importance of space constraints in cities and the uneven spatial distribution of founders of new firms, among other things. These findings have implications for policy in that they define the context in which policy is operating and offer some guidance as to the likely effectiveness of alternative initiatives. But a wide range of policy responses to this new information is still possible. We have recommended that to help overcome physical obstacles to industrial growth in cities, urban local authorities should get together with the major companies in their areas to plan for land requirements for future expansion. An alternative response to the same findings would be to accept the space constraints on industry in cities as given, and to advocate a policy to replace industrial jobs with service jobs, or to plan a reduction of the working population in inner city areas. Similarly, a shortage of founders of new firms in an area could lead to the recommendation that policies are needed to increase the supply, or alternatively that greater efforts should be made to attract existing firms from other areas to make good the deficiency.

Policy formulation is such an inherently political matter that policy recommendations should be clearly distinguished from research findings, possibly in separate publications. This is not always easy because civil servants usually insist that government-sponsored research includes extensive recommendations. Each research finding should nevertheless do something to improve the quality of decision making by widening the range of alternative courses of action or by specifying their consequences. For this reason the results of research are usually more interesting than the policy recommendations, which often deserve healthy scepticism.

5
Editorial introduction

This chapter starts from the position that it is impossible to separate off the question of 'where' employment change occurs from the questions of 'how and why' – in other words that it is impossible to separate geography from production. This is demonstrated in a number of ways. Perhaps most importantly in this paper three different forms of production reorganization are identified, each of which may result in job loss but each of which has different implications for the geography of job loss. Intensification, rationalization and technical change are each argued to have, integral to their definition as forms of production change, different spatial implications. These different geographical implications concern the potential degree of variability of job loss between areas, the likelihood or otherwise of 'location factors' being important at all, and even the role of any particular location factor. In other words, it is argued, geographical pattern may be at least as much an outcome of what is going on in production as of any variation in the characteristics of areas and, further, those very characteristics of areas (location factors) may operate differently (have a different impact on location) depending, once again, on what is going on in production. Finally on this subject it is argued that geography itself may affect the kinds of changes adopted in production. The two – geography and production – are genuinely interlinked, and the one cannot really be understood without the other.

Moreover, the kind of production being investigated is capitalist production and, therefore, in order to understand how it works and why it produces specific geographical outcomes it is necessary to specify it *as* capitalist production, in terms of both its social relations and its underlying dynamic. This is not a question of where you break into the system, but of how production is conceptualized. It means, it is argued, both recognizing the social conflict at the heart of capitalist production and, in particular, distinguishing between production of output and production of profit. It also means recognizing the long-term shifts within capitalist economies. As in the case of chapter 2, however, which lays more emphasis at this last level, these underlying changes or tendencies are not expected to operate mechanistically. While there are underlying causes, there is also great variability of outcome.

This point is important, for it distinguishes the notion of causality (and of its demonstration through research) in this chapter (and in chapters 2 and 6) from that in chapter 4. In this chapter, the different mechanisms of production change are argued to have different spatial implications. But these implications are not discovered (nor are they 'proven') through large scale empirical analysis. Like the underlying processes in chapter 2, they are unearthed through detailed and in-depth conceptual and empirical study. In the case of the implications in the present chapter, they are integral to the definition of the forms of production change. They are *necessary* possibilities. But they are only necessary to those mechanisms in their pure form. In any actual situation they will be operating in and combined with sets of other processes, each of which will affect the operation of the other. The necessary geographical implications are not therefore to be expected to be always directly empirically observable as outcomes. They do not generate hypotheses testable through generality of occurrence and large-scale empirical verification. Common underlying causal structures when operating in the infinite variety of the real world can only be expected to produce an infinite variety of outcomes.

Finally, while the mechanisms of production change may

influence the geography of employment change, they are clearly not the same as the causes of that employment change. Intensification, rationalization and technical change in a particular country or region are themselves only strategies adopted in response to, and as part of, wider changes in the development and organization of the capitalist economy, whether that be at national or international level.

This approach has a number of direct policy implications. It means it may be necessary to intervene in production in order to influence geography. Such intervention may be at national level, local level or at the level of the individual plant. And since it is specifically capitalist production which is under scrutiny, it may also be necessary to challenge that. This does not mean empty advocation of 'changing the system'; it means simply a range of policies each of which in different ways attempts to exert greater social control over production. The fact that the method enquires also into the internal organization of production means, too, that questions are raised which go beyond policy concerned only with numbers of jobs. In particular, it raises issues to do with job quality, skill, job control, and so forth. It raises issues of what any 'regional policy' should be *for*, of what geographical inequality *is*.

Further, accepting as integral to the formulation of method that capitalist production is structured around a social conflict, also means recognizing that there are different policy interests involved. The irrationalities for workers of intensification of work for those with jobs, but in a sea of high unemployment, are not at all irrationalities for capital. Policies are therefore necessarily political and, moreover, are not confined to capital and Whitehall. The chapter therefore also begins to explore some of the possibilities (and problems) of 'policies' for labour itself.

5

DOREEN MASSEY &
RICHARD MEEGAN

Profits and job loss

Introduction

The title of our recent book on job loss was originally 'The Geography of Employment Decline'. On further thought, however, it was finally changed to 'The Anatomy of Job Loss' (Massey and Meegan 1982). This was a significant change, for the basic argument in the book is that it is not possible really to understand the geographical pattern of employment decline without going behind this pattern to an understanding of the underlying structure and mechanisms of the processes creating job loss – hence the also carefully selected sub-title: 'the how, why and where of employment decline'. To explain 'where' you must also be able to understand and explain 'why' and 'how'. And the analytical framework and methodology we adopted to study the geography of job loss were designed precisely to make these connections.

In this chapter, we want to explore the implications of our approach to the analysis of the geography of job loss specifically for policy formulation. The main features of our approach are described in the first section while our recommendations for policy are outlined in the subsequent section. The concluding section explores the relationship between the two.

Aspects of method

To understand the geography of job loss it is necessary also to understand the reasons why, and the ways in which, employment decline occurs. The basic argument of our work on industrial location and changing regional patterns of employment is that *geography and production are inextricably connected. And our analytical framework and methodology both try to recognize this interconnection and explicitly bring into consideration what actually goes on inside the workplace in terms of the organization of production*.

A commonly adopted approach to the geographical analysis of job loss is to confront two patterns: on the one hand, employment change and, on the other, area characteristics or 'location factors'. A high and significant correlation (whether identified in a formal statistical sense or in some other way) is taken to indicate a cause or explanation. What this approach lacks is any notion of process underlying the relationship. The process in question is production: jobs change as production develops and production, as it develops, makes use of, and helps shape, area characteristics.

Bringing production into consideration has a number of important implications for the analysis of job loss. Let us look at the 'employment change' side of the correlation approach described above. This approach implicitly accepts that job loss everywhere is very much the same phenomenon. But is it? We were able to identify three very distinct mechanisms of production change that were important in the job loss taking place in a range of manufacturing industries in the late 1960s/early 1970s.

With *intensification* the employment decline was associated with the reorganization of existing production processes without either loss of capacity or any major new investment being undertaken. Reorganization of this type is geared towards increasing labour productivity within an existing production technique, and the aim is to ensure that the individual workers who retain their jobs produce more in a given amount of time. Intensification may take the form of minor mechanization, exhortation and incentives to work harder, and the physical speed-up of flow-line production equipment, and it may go by such names as 'increased flexibility', 'reductions in overmanning' and, a recent variant on the railways, 'flexible rostering'. In complete contrast, job loss in another group of industries we studied was associated with heavy net capital investment, often related to changes between production techniques and generally resulting in a substantial reduction in the amount of labour required for any given level of output. Employment decline in industries experiencing such *investment and technical change* has tended to be popularly labelled as 'technological' or 'automatic unemployment' even though it is not the 'technology' itself that causes the employment decline, but the combination of the different levels of output and labour productivity change with which it is associated. The other form of production reorganization we identified, *rationalization*, was different again. Here the strategy involves disinvestment and cutbacks in capacity. Production and labour processes remain unchanged, the only alteration being that of scale as the productive base of the sectors affected is pared away.

Intensification, rationalization and investment and technical change, then, are clearly very different forms of production reorganization. What they have in common, however, is the fact that they are all mechanisms through which job loss occurs. Moreover, each has very different implications for the *geography* of this employment decline. Understanding how this argument was constructed, and the status of each of the steps in it, is very important in understanding our overall analytical and methodological approach, so it is worth briefly going over it here.

Our argument is that *different possible geographical implications are integral to the definition of each of these different forms of production reorganization in their 'pure form'.* Thus, the impact of intensification is confined to the existing geographical distribution of factories; there is no plant closure, no significant new investment and no expansion of employment at individual sites; the process does not generate any 'potentially mobile employment' (i.e. the possibility of some locations gaining employment even in an overall context of sectoral employment decline). Rationalization does not involve new investment either, and again, all changes in employment are confined to the existing geography of plants. No new locations are required. But, in contrast with intensification, there may be complete closure of factories, and there may also be some, admittedly limited, mobile employment, where the reorganization involves concentration of capacity at a smaller number of larger sites. This mobility is of course only between existing sites. Taken overall, then, there is a greater possibility of geographical variability between locations with rationalization than there is with intensification. While, with intensification, no location could expect to gain employment, no location will lose all its employment. With rationalization, in contrast, some locations could lose all their jobs while others could actually gain jobs from whatever potentially mobile employment is generated.

Technical change is different again. It always involves the generation of potentially mobile employment for, by definition, there has to be investment in new productive capacity. Moreover, this is new investment and the chosen location could well be a greenfield site. This potentially mobile employment is 'fully mobile' and consequently the employment change when technical change is dominant is not limited to the existing geographical distribution. Thus it is possible to find overall sectoral employment decline linked with quite considerable job mobility. Moreover, since what we are talking about is technical change in the context of aggregate job loss, the new investment in the new production techniques must always, to some extent, be compensated for by some cutbacks and closure of capacity. So, in principle, there are always two aspects to the process, two decisions to be made: where to put the new capacity, and where to close the old. Both may occur on the same site, or they may involve completely different locations. What all this means is that the potential variation in regional/locational employment change is greater again with technical change, because the scenarios range from total closure in one location to the establishment of brand new sites of production.

We examined the regional employment data for the sectors we studied and did find fairly clear differences in the geography of decline between the three groups of industries, differences which broadly coincided with those we had postulated. Net regional employment gains (an admittedly crude surrogate for the existence of potentially mobile employment) was relatively more important in the group of industries dominated by technical change, and the range and variability of regional employment change was greater in the technical change and rationalization industries than in those in which intensification had been important. And, at regional level, the only instance of the generation of totally new employment occurred in a sector in which investment in technical change was taking place.

It should be stressed, however, that *in no sense did we regard these statistical findings as providing a confirmatory test of spatial 'hypotheses'*. The possible geographical implications of our different forms of production change were not hypotheses at all. They were integral to the definition of the forms of production change. Moreover, *it would be wrong to expect these mechanisms to operate in any actual case either alone or in pure form*. It would be very rare for only one type of production reorganization to appear in a sector. We identified the important one, but others were usually taking place at the same time, and the different forms of production reorganization would therefore interlock and affect each other. And there is the added possibility of variation in the operation of the processes themselves. The implications we derived are only necessary possibilities, they are not necessities. With rationalization, for example, there may be no concentration in the reorganization of capacity. Again, while technical change may lead to complete changes of location (i.e. closure and the opening of a greenfield site) it may, in contrast, also result solely in a reduction of jobs on existing sites.

It is for these reasons, then, that we did not regard our definitional implications as being simply predictive. What the exercise did provide, however, was a first way in to understanding what was going on behind the geographical pattern of employment decline in the industries we studied. It certainly cast new light on what job loss, and its geography, actually constitute. The immediate causes of employment decline may vary widely, and so may the geographical implications of those causes. 'Job loss', in other words, is not an adequate theoretical category if what is at issue is its causes and geography. It is a 'chaotic conception' – it combines the causally unrelated.

The approaches which deal only in aggregate numbers fail to recognize this. For under the broad title of 'job loss', the outcome of very different processes may be being lumped together in statistical comparisons. But if it is *explanation* of the geography of job loss that is being sought, it is pointless to look simply at descriptive patterns. It is also necessary to bring in the process behind them – production.

Only by recognizing that production is a *social process* is it possible to clarify the real economic, and political, significance of employment decline. While job loss is always a loss for labour, it is not always one for capital. The occurrence of job loss can in no sense be automatically equated with stagnation or downturn in accumulation. Again this was clearly demonstrated by the differences in the forms of production reorganization we identified.

Neither intensification nor technical change in themselves involve closure of capacity or cutbacks in production. Both of them, by lowering costs, by increasing labour productivity and, for technical change, by changing the product are simply means of increasing or maintaining competitiveness, of *carrying on accumulating*. With rationalization, however, employment decline does go along with disinvestment and cutbacks in capacity, the end result being, as already argued, a reduction in the productive base of the industries affected. But we need to be careful even here about what we mean by 'decline'. The general reason for rationalization is lack of profitability but this does not necessarily mean that losses are being incurred. Profitability is a relative measure: profits may still be being made but remain low in comparison with those in other sectors. Moreover, what is an adequate rate of profit can vary widely between sectors and, within sectors, between companies. Thus, for example, small companies may go to the wall when profits fall but equally, in certain circumstances, small companies may be willing to hang on either because the profits still being earned are 'adequate' or because such companies lack the financial strength and/or corporate horizons to shift into other activities. In contrast, large companies may be better able to withstand losses than small ones. But again, equally, large companies may be prepared to shift their investments between sectors in response to the slightest shift in the rate of profit. *There are no 'rules' of company behaviour*. The point is that the implications of rationalization for individual firms can be very different. It may imply company failure but it may also just be a sign of a company shifting investment into other sectors with higher levels of profit, and by so doing, raising the overall profitability of its operations.

It is clearly not possible, therefore, to assume that job loss in individual sectors, companies and/or factories necessarily implies failure in terms of accumulation. Only when the processes going on behind the job loss have been identified is it possible to draw any conclusions. And, it is not difficult to see what this means for regional analysis. Job loss in one region, for example, may predominantly be the result of rationalization, while that in another may be caused by major new investment and technical change in production. Thus, while both regions are suffering from employment decline, the implications of the two processes behind the decline are very different for their relative economic 'health'.

Our approach thus recognizes that the production we are looking at is capitalist: it is about capital accumulation. This is important both for analysis and, as we shall see later, policy. For there are two sides to capitalist production: the process of production of profit and the process of production of physical goods. It is the former which determines the latter, and by recognizing this the analysis also makes clear the non-complementarity of interests between capital and labour. Conflict is inherent in their structural relationship. All the forms of production we identified were strategies resulting from, and geared towards, the production of profit. And employment decline – always a loss for labour – was an essential part of those strategies. Production for profit meant job loss, and the geography of that job loss was similarly related to the requirements of production for profit.

This does not mean that space was purely passive in its relationship to production change – that it was simply on the receiving-end of aspatial pressures. Integral to our approach is the recognition that geographical characteristics (including, for example, physical geography and such labour characteristics as availability and cost) may themselves influence the operation of these pressures and the form of response of individual companies. Thus, for example, companies able to exploit geographical differences in labour will be better placed to withstand international and sectoral-level pressures to cut costs. And the existence of such geographical differences may encourage the adoption of particular forms of production change. *Locational strategy is thus not simply a result of, it is also a crucial element in, the conflict between capital and labour.*

Further, if production is really to be understood as a social process, that means taking seriously the fact that *the changes we study are neither inevitable nor mechanistically produced.* In our analyses of employment change we have tried to avoid a number of different kinds of imposition of a priori expectations on empirical analysis of particular situations. We have already referred to one way in which we did this. As we have said, an initial step in our analysis was to derive from each form of production reorganization its implications for the geographical pattern of employment change. These were inherent, necessary implications of 'pure forms'. But the real situations which we were examining were far more complex than this. The richness of these contingent conditions could only be expected to have its impact on the actual outcome. This means, first, that there would be no validity in 'testing' the inherent implications of the different forms of production reorganization by the use of empirical statistics relating to directly-observable outcomes. It also means, secondly, that any thorough explanatory empirical analysis must use and combine both an understanding of the necessary relations inherent in the processes under study and the full measure of the specificity of the particular situation in which those processes are operating.

There are other kinds of 'a priori' conceptualization, perhaps more common in radical analyses, which we have also tried to treat with a degree of circumspection. One of these concerns 'tendencies' or 'laws' of capitalist development. The labour process, for instance, has undoubtedly undergone a series of major changes which have been systematically related to wider developments in capitalist production. These major changes are important to understand and it is important also to relate shifts in the work process within individual firms to them. What we would reject, on the other hand, is attempts to *explain* changes at that level simply by reference to 'tendencies' resulting from the 'inexorable logic of capital accumulation'. Similarly, we do not believe that the behaviour of individual firms can be explained in terms of the 'requirements of accumulation'. We really mean that production is a social process. The way in which a particular firm responds to the wider pressures upon it will depend on a wide range of factors including in particular the social nature of the capital involved and the organizational capacities of labour. What is important is to steer a path which encompasses both a recognition that actual causes take place through detailed and specific social mechanisms and an appreciation of how this detail relates to the wider context of production in a capitalist society.

This is to focus on company level. There is, of course, also 'room for manoeuvre' at the wider economic and political level. Our analysis at company and sectoral level allows us to bring in wider causes in the national and international political economy, and these wider causes are not just to be found in the realm of the purely 'economic'. Our analysis of production reorganization at the level of sectors and companies also fits in with a broader analysis of the political climate in which those reorganizations are occurring. Thus we were able to link the dominance of specific forms of production reorganization in particular periods with broader political and economic conditions. Technical change was far more important in the late 1960s/early 1970s than it has been under monetarism, when rationalization and, to some extent, intensification have become more dominant. And this change in dominance can be related to changes in both economic and political circumstances. It is not just that economic conditions were very different in the earlier period. The political interpretation of those conditions and the policies consequently adopted were also markedly different from what they have been since the onset of monetarism. We will return to this point in the next section when we discuss policy conclusions and recommendations.

All this raises a number of more day-to-day difficulties for analysis. It means being aware both of what is happening in individual companies and of the wider economic and political situation. We do feel that the use of Census data is not enough and that it is also important to use interviews and to draw upon as wide and varied a set of data and information sources as possible. 'Systematic and complete coverage' is not the only criterion for a good source of information. The processes we are trying to understand are far more nuanced, richer and more complex than that. Our approach also raises questions about the social sources of data. Most of the sources of information about changes in the labour process are produced by organizations representative of management and/or the state. Information is thus more readily available from one side of what is essentially a conflictual social process. Our experience is that it is important to talk to those on the other side of that process – trade-unionists, people on the shop-floor, etc. – as well. The job of the researcher is then to structure together what are often fundamentally different understandings of what is going on.

Policy and politics

The traditional instrument for central government intervention in the geography of employment change has been regional policy. Our methodology and findings have a number of implications for assessing the current relevance and likely effectiveness of this policy. It is commonly argued that the existence of widespread job loss has resulted in a diminution in both the scope and effectiveness of traditional regional policy. The argument clearly has some force, but our findings show that things are more complicated than this. The political argument usually put forward is that it is not possible to have regional policy when employment overall is declining – because there is no mobile employment to influence. There is, of course, less mobile employment, but it is not true that there is none. The second half of the 1960s and early 1970s was a period in which manufacturing employment was declining overall but when, even so, there was plenty of potentially mobile employment around to be influenced. Our analysis throws light on why this was so, for it demonstrates that the amount of potentially mobile employment will vary depending on how job loss takes place. It is not only decline itself, but also the mechanisms of that decline, which are important. And, as already pointed out on p. 127, the balance of these mechanisms has changed over the past decade or so – with significant implications for the scope and effectiveness of regional policy.

The second half of the 1960s and early 1970s was the high point of regional policy. It was certainly a very distinctive period in the recent economic history of the UK. Whilst clearly a time of relative national economic decline, there was at least some growth going on, both nationally within particular industries, and internationally. Moreover, while manufacturing employment fell over the period it did so in the context of output growth and, therefore, substantial increases in labour productivity. Indeed these productivity increases finally reached those of the UK's European competitors in this period (see, for example, Jones 1976). The politics of the Labour Government of the day were to modernize the productive base through sectoral restructuring, productivity bargaining, and 'the white heat of the technological revolution'. We would characterize this period as one of 'active decline'. Technical change in production was being introduced and there was therefore plenty of potentially mobile employment on which regional policy could act. And this aspect of what was going on in production was one of the important conditions of the effectiveness of regional policy at that time. Since that period the context in which regional policy has had to operate has been dramatically changed. Manufacturing employment decline has accelerated and spread both to more and more industries and to more and more regions. And behind this accelerated decline lies a marked change in economic and political conditions. Since 1973, the UK's economic problems have been set against a background of decline in the international economy. And, especially since the election of 1979, the policies towards industry have changed. The earlier emphasis on modernization and intervention has been replaced by a politics explicitly directed towards shifting the balance of power between capital and labour. The stumbling and often misplaced attempts at achieving a consensus in the earlier period have now been replaced by straightforward confrontation. 'Wilsonism' has been followed by 'Thatcherism'.

Employment decline today is very different from that which took place in the late 1960s/early 1970s. In particular, the balance between the different mechanisms has changed.

Rationalization now appears to be a far more important cause of job loss than it was in the late 1960s/early 1970s – as the collapse in manufacturing output and the dramatic increase in bankruptcies and company liquidations testify. Follow-up studies of the industries we looked at that had been trying to maintain their competitiveness through intensification or technical change showed that most were now rationalizing. And the very fact of widespread rationalization has facilitated the introduction of measures to intensify the work process. In marked contrast, aggregate data on investment patterns and research and development expenditure, and detailed industry and company studies all seem to point to technical change now being far less important as a source of job loss than it was in the heyday of the 'white heat'. Where productivity increases are occurring they are being achieved predominantly by differential rationalization and/or intensification.

The implications for regional policy of this shift in balance between the different mechanisms of employment decline are clear. Most importantly, the amount of potentially mobile employment is now likely to be far lower tha it was in the late 1960s/early 1970s, as a consequence of the relative decline in importance of major greenfield investment in technical change. With rationalization, individual points of employment growth are related to concentrations of existing capacity on existing sites, rather than to the establishment of new locations, while with intensification there is little likelihood of any mobile employment at all. The ability of central government to have any social impact on the geographical distribution of employment, even if it wanted to do so, through the old kinds of regional policy based on incentives to mobile investment is now clearly extremely limited.

This is in no sense meant to be a nostalgic argument for traditional regional policy. That policy certainly had its achievements – its reinforcement of the trend towards the provision of waged work for women in the peripheral areas being possibly the most noteworthy. But its operation was also part of the cause of the accentuation of a new form of regional inequality, based not simply on differences in levels of unemployment but also on differences in the types of job available. The continuing decline of skilled manual work for men in the peripheral areas contrasted with the continuing concentration of 'higher-level' administrative, professional and scientific jobs in the South-East. Repetitive, semi-skilled and low-paid work in manufacturing and 'low-level' clerical jobs, both increasingly employing women, grew in the Development Areas. 'Branch-plant economies' became an important feature of this new geography of social class. Thus, not only was regional policy's success dependent on what was happening in production at that period, so also was its failure. On the one hand its ability to influence the geography of jobs was enhanced both by the availability of potentially mobile employment and by changes in the labour process. On the other hand the nature of these changes, the emerging forms of 'new technology', led to a division of labour within production which was the social basis for the new spatial division of labour.

The first, and central, element of any new policy proposals to influence the geography of jobs must therefore be that such *policies must be integrally related to strategies for the economy as a whole, for the organization of production and for technology*. This does not mean simply that it is necessary to call for national economic growth before the regional problem can be solved, though we would agree with that. It also means that planning at national level should itself take account of geography. It is not a question first of getting national growth and then thinking about its spatial form; the two are integrally related. But it also means thinking about the internal organization of production: the quality of jobs, the geography of ownership and real control, the social form of technology.

This brings us back to the question of what we actually mean by 'production'. As we have stressed, production in a capitalist society is production for private profit. It is geared not towards the production of goods for their use value but towards the possibilities of financial profits that their production offers. All the job loss we studied, and its geography, went back to strategies and calculations based on the requirements of production for profit. If a serious challenge to these kinds of employment decline is to be made then a serious challenge will also need to be raised to the social relations of production for profit. If there is a basic policy message in our research then this is it. *To combat these kinds of job loss and to influence the geography of employment, it is necessary to exert greater social control over the organization of production*.

This can take a number of forms. Each of the different kinds of production reorganization we identified raises different issues concerning the social organization of production and provides different opportunities for political action and policy development.

Intensification raises issues of control over the work process, over established rules and procedures, over the organization and quality of daily life in the workplace. It may involve speed-up, increased flexibility, new disciplinary procedures, shorter tea-breaks, and flexible rostering – all in the name of improving competitiveness and profitability. As unemployment mounts more and more jobs are lost as individual workers, those fortunate enough to retain their jobs, are made to work harder and harder. In any terms other than those of the calculation of profit such a process is surely irrational. It is our contention that a politics to combat job losses which are resulting from intensification should highlight this irrationality. Instead of simply subsidizing job retention as such, or going on the defensive just on the issue of the numbers of jobs, the issue could be broadened to relate to the actual *process* which is going on – to raise issues of the quality of, and control over, working life in the office or factory. This means going on the offensive, to politicize the economic and social issues that intensification raises. In the 1960s and 1970s the clothing industry must surely have offered opportunities for such a campaign as job losses, speed-up, and the growth of a new sweated sector went hand in hand.

Like intensification, technical change raises important political questions about the organization of the work process. Here again, the urgency of the immediate need to defend jobs often forces workers and unions into a stance which seems resistant to technological change itself. But the real issue is not whether to have technological change or not, but the nature of that change and its wider relation to production. Technical change can eliminate boring and dangerous jobs, or it can create mindless and repetitive ones. It can create unemployment, or lead to new job opportunities. There are no technologically determined paths to follow; choices can be made. The aim of much technological change in a capitalist society is not to make the work process more humane, but to cut labour costs or increase managerial (and decrease employee) control. This affects the division of labour. On the one hand a scientific élite has been established, laying claim to possession of increasingly rarified technological information and knowledge; on the other hand other jobs are de-skilled. To challenge the nature of technology it is also necessary to challenge the control of it. Production strategies need to be derived based around 'alternative technology'. Basic to these strategies is the development of 'human-centred' production systems, which start from the premiss that technology need not be de-skilling and that processes can be devised which depend upon, and interact with, human judgement and skill. This in turn may involve challenging profit as the only possible criterion of production.

The operation of the 'white-heat' technology policy has shown the dangers of a strategy which *fails* to question fundamentally the nature of the technological changes it promotes. In this case, the promotion of mass-production techniques and the encouragement of centralized research and development facilities – a hierarchical organization of production which, in a sense, mirrored the 'top-down' nature of the policy. By encouraging this particular form of technological change it also encouraged the development of the new forms of geographical differentiation we mentioned earlier, with the concentration of research and development within the South-East.

Many of the most effective policies for 'new technology', which have begun to challenge the social, and geographical, concentration of technological development, have been built around local initiatives. Holland's 'science-shops', deliberately aimed at encouraging local involvement in, and development of, 'alternative technologies', are an important step in this direction. Again the Greater London Enterprise Board is setting up 'technology networks' throughout London. Workshops are being established drawing on the scientific and engineering facilities of universities and polytechnics and providing facilities for trade-unionists and other groups in the local area to participate in the development of new products using the kind of 'human-centred' production technologies already referred to. Such initiatives are a radical attempt at both an economic and a social regeneration of the areas in which they are located.

With both intensification and technical change, the main political issues revolve around the control, and nature, of production – *how* production should be carried on. Rationalization, in contrast, raises the fundamental question of *why* production should be undertaken in the first place. Under a capitalist formula goods are produced not for their use-value but for their exchange-value, for the potential they offer for making profit.

On the whole in Britain, and certainly in the private sector, goods are not produced unless someone can make a profit out of them. This is so, regardless of the fact that many people need those goods, or would like them. In *The Anatomy of Job Loss* we pointed to the apparently patent 'irrationality' of the simultaneous existence at that time of huge stocks of bricks piling up around the country, the hundreds of thousands of unemployed building workers, and the obvious need for new and better housing. Such a situation is irrational if you think in terms of physical assets, of what you could do with the real resources which are available. But it is not necessarily irrational if the calculation is made only in financial terms, in term of profit and loss. It is also, in other words, to the requirement of production for profit that we need to look to understand private decisions to *stop* producing particular physical goods. This is the political question which rationalization raises – *should* production be determined solely by calculations of profit and loss? Different strategies are available depending on considerations of plant / company profitability and the corporate context in which rationalization is introduced. And the different ways of fighting rationalization each raise different political issues.

Thus most 'horror' is expressed where particular plants earmarked for closure can be shown still to be earning profits. This kind of defence has been important in the battles over the closure of steel plants, particularly in the case of Consett and Llanwern. More recently, much press coverage has been given to the 'baffling' decision by Plessey to close its Romford aerospace factory whilst it is still 'making a comfortable profit'. But, as our work on production reorganization illustrated, it is excellent capitalist rationality to close down profitable plants if higher profits can be made elsewhere. Plessey's decision to close down its Romford factory is far less 'baffling' when viewed in the context of the company's current diversification strategy, which involves a shift into the potentially highly profitable area of satellite telecommunications. The trap that these challenges to rationalization fall into, of course, is to accept implicitly that lack of profit is an adequate justification for plant closure – in other words, to argue on capital's terms.

Other policies have been developed, however, which shift the attack by directly challenging the overall rationale of production for profit. One such policy is to take into account the social costs and benefits of plant closure by drawing up social audits or social balance sheets which challenge the private nature of the cost calculations behind plant closure. Against the private savings of closure for the company involved are set such public costs as unemployment benefit for the workers who lose their jobs, lost tax revenue, and multiplier effects on other parts of the local economy. The effects of unemployment on health and the requirements for health care are concerns which are increasingly being taken into account in social audit approaches.

Another policy, developed in the US, focuses on the relation between individual plant profitability and that of the company as a whole. Here, the strategy is to attempt to get the company involved in any closure decision to pay some of the social costs of that closure. This is clearly most applicable to multi-plant companies, which are shifting production around, and which are, overall, still profitable. Penal rating on deserted factories has been one, albeit limited, response to this at the level of local government in the UK. But, in the absence of any national policy, such strategies have had to be tempered by consideration of the impact on the relative 'attractiveness' of the area for other potential inward investment. This constraint has, of course, been partly removed in recent years by the reduction in levels of mobile investment.

All of these social cost approaches to rationalization remain, however, at the level of calculation of individual plants or companies. Another approach is not to argue about profitability at that level at all, but to put the issue at the level of society as a whole, and to argue for socially useful production – for the production of physical goods that meet real social needs rather than private profit. Here the attempt is to change the level at which financial calculation takes place. The economy as a whole may need to 'show a profit' but this need not apply to each individual point of production. This again, of course, challenges the private ownership of production, but it is not only in the privately owned sectors of the economy that the argument applies. For in so-called mixed economies private-sector rules may often apply to public-sector investment. This is the crux of the battle over pit closures in the coal industry in Britain in the early 1980s. The Coal Board argues on profit and loss terms, pit by pit. The mining union, in contrast, replies that such considerations are not the only ones and may anyway be very short-sighted. A colliery once closed can rarely be reopened, and to close a colliery before its reserves of coal have been exhausted is therefore to lock up for ever, effectively to squander, a real physical resource.

Others, in very different parts of the economy, have also tried to challenge the notion that immediate and individual plant or company profit should be the only basis for production. The Lucas Aerospace workers' alternative plan has shown the way with its proposals for redirecting production towards products designed so as not to waste energy or raw materials, and to be produced by labour-intensive and non-alienating work forms (Wainwright and Elliott 1982). Other groups of workers have followed this initiative and developed their own strategies for alternative production. Whilst the Lucas Aerospace plan was initially conceived in a context of sectoral and company growth, these new alternative plans are now being drawn up in sectors, companies and areas severely affected by job loss (Labour Research Department 1984).

Plans for 'alternative production' to combat job loss are now being developed by local enterprise boards which have been established by some progressive local authorities and charged with powers to intervene in the organization of production in their areas.

Their policies are all at early stages but the possibilities they offer for policy formulation are becoming clear. First, they are attempting to be non-competitive between areas. Second, while, for legal reasons, they have had to take into account commercial considerations in their interventions, they are developing an overall social-cost framework for their long-term investment packages. Third, their investments are combined with Planning Agreements, recognition of trade unions, and undertakings over work conditions. They are looking seriously at job type as well as job numbers and it is in this context that the policies for alternative technologies are also important. Fourth, they explicitly encourage employment initiatives from a wide range of groups in the local community and, together, are building up a network for the exchange of experience in policy formulation. They are thus providing the building blocks upon which any future national policy for employment decline could and should be erected. And, last but not least, this local intervention offers the genuine possibility, both of building policies which are tailored to the needs of particular areas, and of removing the fatalism and defeatism engendered in local communities by past job loss associated with the kinds of production reorganization we have discussed in this paper (Blunkett and Green 1983).

All these arguments which challenge production for profit and which involve Exchequer and local government subsidization of socially useful/alternative production might seem to some as pie-in-the-sky romanticism. Yet firms are already subsidized, in some cases quite heavily. The rationale may be that jobs are saved or depressed areas helped. The point, however, is that these subsidized companies remain structured towards profitability. Subsidies are necessary for profits to be made. Where profits *are* made these go, of course, to the private owners of the subsidized companies. Profits may equally not be made, even with subsidy, and the loss is often borne by the taxpayer. The De Lorean episode is a scandalous example of a simple 'subsidize–profits' approach to industrial and regional policy. And there are many more respectable companies which interact with government policy in a similar way. The question that needs to be asked in this context is, why should not the same amounts of money be spent as part of a policy which recognizes from the beginning that production may not make a profit and which deliberately chooses to support, not the production of cars for the extremely rich, but socially needed products? Jobs are still preserved. One difference in this scenario is that the subsidy is shifted from the private producer to the consumer. Another is that it changes the criterion for production from private profit to social use. Instead of just being defensive about numbers of jobs, it goes on the offensive about the rationale for production, the reason for production in the first place.

There will, of course, be the usual cries of 'how can decisions be made on social need?' and 'money cannot be poured indefinitely into unprofitable activities'. But both occur already. The current levels of defence expenditure and 'rationalization' of the National Health Service, for example, have been deemed to be 'socially necessary'. And, as a recent report of the Energy Committee of the House of Commons made clear, the economic case for nuclear power has not been made. Yet ever-increasing amounts of money are being pumped into the nuclear programme. The cost of the Sizewell B reactor alone has been put at £1147 million (in 1982 prices). In 1982 the economy and resources were mobilized to defend a small group of unheard-of islands lying in the cold mists off Cape Horn. Could not something like the same effort and commitment be put into mobilizing productive resources for social need?

Politics and method: some connections

What then is the relationship between our analytical and methodological approach and our policy conclusions and recommendations? We hope to be able to demonstrate how the approach we adopt helps to frame both the scope and nature of the policy conclusions outlined in the previous section. Whilst we would not argue that our approach determines our policy recommendations the two are definitely related.

Central to our approach is the notion that the geography of jobs is inextricably interlinked with considerations of what is happening in production. By making this connection in our analysis we avoid coming up with policies aimed exclusively at influencing the geographical distribution of locational characteristics or factors – a trap into which the 'correlation approach' easily falls. If you are correlating a list of locational factors with regional employment change and a high R^2 turns up, apparently indicating the significance of premises (size of, or lack of) and/or infrastructure, for example, then policy will tend almost automatically to focus on the winning factor(s): 'build more, or bigger premises', 'improve the road system', and so on. Location factors thus become the 'explanation' for the employment change. And it is not difficult to see how the methodology itself encourages this. Yet it is dangerously misleading. We would argue that whether or not location characteristics are important *at all* will in part relate to the kind of production change going on. With technical change, there is always some, at least implicit, locational decision to be made, concerning the location of the new investment – on site or not. With rationalization and intensification, location factors, in the sense of the more general characteristics of an area at least, may be important in some cases but equally may also be completely irrelevant. We found a number of examples in our detailed case studies where locational characteristics played precisely no part in the geography of the production reorganization. Characteristics of the firm and/or production process in question proved to be more important than regional or locational characteristics. We were also able to show how the same locational factors or plant characteristics may operate in very different ways depending on the kind of production change going on. Differing levels of worker militancy, for example, are often quoted as reasons for differences in job loss by region. But this 'factor' clearly operates in different ways according to the different forms of production reorganization we identified. Where intensification is going on, it may mean that a well-organized workforce is better able to resist job cuts, so less jobs are lost in the militant regions than in other less-organized places. With rationalization, worker militancy might actually precipitate job loss with the 'dispute-prone' plants being singled out for closure. In cases of technical change, the same 'factor' might influence the location of the new capacity and any closure of old. Thus more 'militant' areas might lose jobs, while 'less organized' workers elsewhere may gain. What this means, of course, is that at any given time the same locational factor may be operating differently with different kinds of production reorganization. To combine them in aggregate figures of job loss and search for common factors would bury such differences.

While a location factor may well be important as part of an explanation for regional employment change, in itself it cannot be the explanation. *It is processes not factors which are explanations*, and location factors only become important when considered in the wider context of the production process. Adopting an approach which recognizes this avoids that tendency to blame the victim, which approaches based on area characteristics so often slide into. Shift-share analysis, by identifying differential components which supposedly relate to the specifically regional element of performance can reinforce this tendency, as can any formulation in terms of 'regional performance'. Too often it is argued that jobs are disappearing because the regions affected are 'deficient' in terms of one or other of the significant location factors. In such an analysis, the 'fault' thus lies with the regions. The inner city debate of recent years has shown where this kind of analysis ultimately leads. High levels of job loss in the inner cities are blamed on the fact that the latter *are* inner cities. But this is to forget history, and to ignore process. It was the development of production that favoured employment growth in areas that are now cities, and it is now abundantly clear that production no longer favours city locations. But if analysis is to avoid the real danger of confusing the *spatial location* of the operation of processes with the processes themselves – of mixing up effect with cause – it is necessary for it to examine the development of the process of production itself and, in so doing, to unravel the mixture of production and geographical factors which combine to influence this development.

We are not arguing that all proponents of the correlation-type approach ignore production. On the contrary, having found a factor 'with a high R^2', many go to great lengths to link the factor to production. But herein lies another trap – the tendency to accept the rationale of production for profit. The existence, or absence, of certain local characteristics means that the location in question is not providing the right conditions for profitable production and the policy implication usually drawn from this is that those characteristics must be provided. The argument in other words, is 'change a local deficiency in terms of the latest location factor into an "incentive"' – build a science park, say – and jobs will follow. If mass-production factories are what capital wants, provide the site, the buildings, the grants – people will just have to learn to love them. *Thus policy starts from capital's requirements and attempts to adapt the region and its inhabitants to them. The alternative approach is to start from the requirements of the people in the region and to adapt production to these, to accept that the real problem is that production as it is presently organized is not providing enough jobs to go around.* Is it the inner cities which have failed capitalist production, or capitalist production which has failed the inner cities?

Production is a social process and, in capitalist production, inherently involves conflict between the interests of capital and those of labour. As we were careful to argue earlier a loss for labour is not necessarily one for capital. So if employment decline is seriously to be combatted then awkward questions have to be asked and radical policy decisions have to be made. Propping up profits with subsidy is not necessarily propping up jobs. It is essential for policy formulation to go beyond the usual formulation of the 'choice' between, for example, subsidising BL's attempts to make profits (for whatever jobs these attempts require) and letting it 'fail'. The failure here is the failure of the company to make profits, not its inability to manufacture cars. Which brings us back to the dichotomy, in capitalist production, between the production of physical goods and the production of profit – use-value versus exchange-value.

All the forms of production reorganization we identified, and the job loss associated with them, were the result of production being geared towards the production of profit. To challenge the job loss it is therefore necessary to challenge this profit orientation. This is why we do not argue, say, for the continuance of a subsidy policy structured towards making companies profitable, even though this would involve saving some jobs – a profits subsidy dressed up as a labour subsidy. The organization of production would remain unchanged. We are arguing for a financial policy geared towards restructuring industry in the direction of socially useful production. This redirection involves questioning why goods are produced and what goods are produced.

It also questions the way in which production is undertaken. Production for profit can lead to intensification and the introduction of alienating technological change. When we say that production is a social process we are not simply referring to its function of collecting people together for the production of goods – whether the latter be for their use-value or their exchange-value. The work process is also about individual and co-operative expression and fulfilment. Social production is not simply concerned with producing goods that are socially needed but also concerned fundamentally with the way in which those goods are produced. An approach which recognizes this concern will naturally frame policy recommendations which go beyond consideration of simple employment numbers, and the geography of that employment, to questions of the type of work and the nature of work relationships. Integral to this policy formulation is examination of the nature of production technologies and labour processes. And this is the reason why we stressed in our policy conclusions the crucial importance of such developments as human-centred manufacturing systems in the formulation and implementation of strategies for alternative production – alternative products produced by alternative processes.

Separating the interests of capital and labour in production also implies widening the range of actors involved in policy making. Too often it is assumed that policy towards industry can consist only of measures to enable Whitehall, or some arm of the local state, to influence management. But there can be strategies for labour, too. Indeed some of the policy recommendations on pp. 130–7 are necessarily addressed to labour (politicizing the various forms of job loss, for instance) and some of the strategies which we advocate have been developed by trade-unionists (alternative plans, for instance). Unlike many other approaches, this approach gives room in analysis for labour to act, and hence allows policy recommendations to be addressed to labour as well as to capital and the state. Indeed, the approach views all three as having 'room to manoeuvre' in their structural relationship – for, as we also stressed earlier, we eschew any mechanistic causality of the 'inexorable logic of capital accumulation' type. If the process was as mechanistic as such approaches would lead us to believe, there would be no such room for manoeuvre, and attempts at policy recommendation would be meaningless.

'Policy making' is thus not seen as being centralized, either at national government level, company head offices or national trade union headquarters. Individuals within factories and offices can also come up with policies that can influence decisions made at 'higher' spatial levels. Much of the disillusionment with traditional regional policy, for example, can be attributed to its failure to allow any genuine local involvement in the planning processs. It was a 'top-down' policy, resulting from an analytical and methodological approach which encouraged a hierarchical view of the world, both socially and geographically. In contrast, our approach allows the formulation of policy which builds upon local initiatives from the 'bottom-up'.

Such strategies for labour are not easy to follow. If your job is threatened the most obvious course of action is simply to defend it, not to raise the wider political issues which its loss involves. Moreover, each of the different mechanisms of job loss which we identified, each of the different forms of production reorganization, can be used in different ways to divide the workforce. Each can be used as a tactic in the conflict between capital and labour. A situation of rationalization leads easily to workers in different factories competing with each other to avoid closure. The recent years of high unemployment have seen numerous examples of this. With technical change, particularly in a period of recession and overall employment decline, the fact that opening a new factory will often entail closing an older one elsewhere can also lead to bitterness and resentment. But as is illustrated in the last section such problems of divisiveness are most apparent when the resistance to job loss remains simply defensive. One way to start overcoming them is precisely to widen the political issues at stake.

But does that not entail us in a contradiction? On the one hand we are arguing that local-level change and job loss can only be understood in a wider context of national and international capitalist production. On the other hand we are arguing that local-level initiatives must be an important element in building a response. There are a number of points here. First, while it is correct that in some sense the 'ultimate' cause of the job losses we have been discussing is a national and international system of capitalist social relations, it is fruitless to look in the short term for major systemic changes at those levels. There will be, neither internationally nor in Britain, no one-off storming of the central citadel after which 'the whole system will be changed'. Nor would that, in a society like that in Britain, necessarily be the most democratic way of going about things. Second, it must none the less be clear from the foregoing analysis that one of the crucial tasks must be for links to be built between those working in different plants within the same company (through combines, for instance) or between those working in the same industry but in different locations, different regions, different countries.

Third, we are arguing, in other words, that the initiative for change must come at *all* levels. It must, as we have said, span the spectrum from including geographical considerations as an integral part of any national economic recovery to building upon alternative plans for social production devised at plant level. Above all, perhaps, it is important to build that social and political confidence which will encourage and enable the development of radical policies which seriously challenge the why, how and where of employment decline.

6
Editorial introduction

Sayer and Morgan begin by recognizing the coexistence of a wide range of groups and interests with a relation to the regional problem, and by establishing as a question to be investigated the nature of that problem. The two things are related: 'The regional problem does not exist in a vacuum, it is a problem *for* particular groups and interests. It may not be a problem at all for some, and others will be affected by it in less evident ways'. It is necessary for the researcher, too, to be clear, therefore, which regional problem it is which he is investigating.

The focus of the empirical work in their own study is a group of individual companies and, as in other chapters, explanation takes place through analysis at two different levels: first at the level of the national and international competitive context and second at the level of the firms themselves. As in other chapters which focus on individual corporations, the relationship between these two levels is crucial. Sayer and Morgan stress, as do other authors, that there is no mechanistic cause from international context, say, to individual firms. The latter may respond in a variety of ways to the former, and both levels, therefore, are necessary for explanation.

The approach adopted is explicitly 'intensive' rather than 'extensive', and the authors present a detailed discussion of the differences between the two approaches, which picks up on many of the themes which have been running throughout this collection. They take up again the issue of the kind of question being asked. For extensive research this will be concerned with the examination of common patterns whereas for intensive research the question concerns how a causal process works out and is structured in actual, particular, cases. This different starting point encapsulates the complete contrast between the two approaches, a contrast which is reflected, as Sayer and Morgan show, in the kinds of groups and categories which are discerned and studied (see chapter 1), in the counterposition between degree of descriptiveness, 'representativeness' and causal explanatory power (see chapter 1, and the contrasting position to this in chapter 4) and in the methods of investigation which are adopted. They also take up again the question of the degree to which the two approaches can be complementary, and point in particular to the importance of the categories adopted in extensive research being related to causal explanation, rather than being simply taxonomic groups based on descriptive similarity. In other words, the argument goes, although extensive research has its functions, for instance in the initial exploration of data, in practice it rarely measures up to the demands for real explanatory power which it would need in order to mesh with intensive analysis.

But if lack of explanatory power is the charge most often levelled at extensive research, what of the 'unrepresentativeness' of which intensive research is accused? As we have seen before (e.g. in chapter 3), the real question concerns how that issue is understood. The underlying causal mechanisms unearthed by intensive research are 'necessary relations or properties of objects' and therefore by definition generalizable to all occurrences of the object to which they refer. But 'actual concrete processes or events are produced through a combination of necessary and contingent relations, . . . the research findings describing these are unlikely to be generalizable to other contexts' (see also, for the same position, chapter 5).

All of this raises questions of engagement. This view of the relations between a company and the wider organization of the economy, although it grants companies and individuals latitude (and therefore responsibility) in the nature of their response, it does not see them as the ultimate cause – 'it is not the individuals but the form of social organization which we would expect to criticize'. In this there is some contrast with the position taken, by and large, in chapter 3. Similarly, if the aim is causal analysis rather than representativeness and replicability, interviews should take advantage of their interactive nature, and should be responsive and flexible. And the recognition that different actors (with different interests) are involved, means that the researcher necessarily has to make judgements of their different explanations. 'Critical evaluation cannot be dispensed with or tacked on as an optional extra under the heading of "policy implications".' Here, too, is a crucial difference from the position adopted in chapter 4.

Such a position means, too, that in direct questions of policy analysis there can be no pretence to an overall view. There are different actors in the game for whom the issues and the range of potential solutions are very different. There are also different kinds of strategy which each of those actors can adopt. Sayer and Morgan close this collection with an analysis of the range of policy prescriptions which this complex set of strategies and actors (including academics) can give rise to. Like other authors they argue the necessity to question the dominant capitalist form of the social relations of production. Like others, too, they recognize the difficulty of doing this, but argue that 'attempts to find alternative ways of defining and calculating economic rationality need to be criticized constructively rather than dismissed'. Behind such debates lie questions even of the definition of a 'solution' – development in, or development of – regions. As they conclude: 'Far from being a wholly academic matter, the question of method is of crucial political importance in generating information that can be socially useful.'

6

ANDREW SAYER &
KEVIN MORGAN

A modern industry in a declining region: links between method, theory and policy

The regional problem does not exist in a vacuum, it is a problem for *particular groups and interests. It may not be a problem at all for some, and others will be affected by it in different, less evident ways.* Suggestions for solutions to the problem must therefore take account of whose problem it is: this question is seldom addressed. We believe that in all empirical research there is a three-way interaction between method, theory and policy conclusions or evaluations. No single one of these is ultimately determinant and they tend to adjust to one another – indeed we have not found it possible to separate them completely. In this chapter we will discuss the interplay between these elements in the context of a research project on electrical engineering in South Wales.[1]

At the outset it should be noted that while our research on this project is relevant to work in geography on regional problems and industrial location, it is intended to be interdisciplinary, investigating aspects which relate to industrial economics and sociology as well. These are some of the issues which prompted us to do the research:

- — the belief that new industries such as electronics could be the salvation of the backward regions, reflected in the idea of 'picking winners' in attracting inward investment
- — the widespread concern about increasing external control and 'branch plant economies'
- — the belief that the instruments of regional policy (e.g. Regional Development Grants) are inappropriate for combatting unemployment in the Development Areas
- — the belief that more generally there are not only conflicts of interests between working communities and local industries but within each of these, between different types of worker and different types of firm.

We wanted to explore these issues more deeply in a particular region, for we felt that the first two needed greater qualification and the last two were under-researched. Above all we wanted to get some insights on how the regional problem persisted despite a significant influx of new industry, how job loss in traditional sectors came to be supplemented by job loss in modern industry, and how working class experience of and responses to working in the region have changed. This then is the context of our research. It is too early yet to present research findings but it may be useful to say something more about the content before discussing the links between method, theory and policy.

Our research on the electrical engineering industry in South Wales includes detailed study of about 25 firms in consumer electronics, domestic electrical appliances and electronic capital equipment (including electronic components, telecommunications and defence equipment). The focus is not on location decisions but on explaining variation in *in situ* performance, particularly as regards the quantity and quality of employment. *The explanation is by reference to two levels, first the competitive context (nationally and internationally) in which the plants developed and, second, the local level in terms of local conditions and what the firms actually did 'on the ground'.* The first level, although frequently omitted in industrial geography, involved 'sectoral analyses (which were intended to discover those parts of the changing economic environment that were specifically relevant to the plants). The second level of research is being done largely by interviews to find out what happened *in* the plants. We believe that *the performance of particular plants cannot simply be 'read off' from a knowledge of sectoral and local conditions*: what happens 'on the ground' also depends on how these resources are organized in the factory and on the kinds of responses to the conditions at both levels, both of which can vary considerably. The second stage of interviewing management, unions and in a subset of cases, workers, will also allow us to learn something of the divergence of interests involved. In general, the focus on particular plants allows us to get beyond the usual practice of seeing them as statistics and in terms of a (managerial) 'view from above'.

Our reasons for choosing electrical engineering were not random. The main factors were as follows:

(i) It has been the most prominent SIC[2] Order in terms of inter-regional industrial movement between 1945 and 1975, and the most important source of employment arising from new openings of plants in Wales in particular.

(ii) Many of its MLHs[3] (especially the electronics-based MLHs as distinct from the electrical MLHs) are in the forefront of advanced product and process technology.

(iii) During the recession since 1979 SIC IX has been most resistant to the slump in terms of output: whereas total manufacturing output declined from 104.7 in 1979 to 89.1 in the second quarter of 1982, SIC IX remained buoyant with output of 113.5 in 1979 compared to 113 in the second quarter of 1982 (1975=100).

(iv) Despite popular impressions (and development agency assumptions) it was a good example of a 'new', dynamic sector that was characterized by jobless growth: total SIC IX employment declined from 745,000 in June 1977 to 617,400 in February 1982 and, although considerable variations exist at MLH level, only two MLHs recorded a higher employment level in 1980 over 1977 (i.e. MLH 363 and MLH 367).

Intensive and extensive research design

In choosing a sector analysis rather than a comprehensive, multi-sectoral/multi-regional survey we have opted for an 'intensive' rather than an 'extensive' research design[4] (see Table 6.1). Superficially this distinction seems nothing more than a question of scale or 'depth versus breadth'. But *the two types of design ask different sorts of question, use different techniques and method and define their objects and boundaries differently*. In intensive research the primary questions concern how some causal process works out in a particular case or limited number of cases, e.g. how was industry restructured in a particular period. Extensive research, which has been far more common in economic geography, is mainly concerned with discovering some of the common properties and general patterns in a population as a whole, e.g. what have been the main changes in the location of industry?[5] Typical methods of extensive research are the large-scale formal questionnaire or interview survey of a 'representative sample' of the population or perhaps the whole population, descriptive and inferential statistics and numerical analysis (e.g. cross-tabulations). Intensive research typically uses less formal, less standardized and more interactive interviews and mainly qualitative forms of analysis.[6]

The approaches also work with different conceptions of *groups*. Extensive research concentrates on *taxonomic* groups,[7] that is groups whose members share similar attributes but which need not actually connect to or interact with one another; e.g. 'plants employing 1000+ employees', or 'free-standing towns' (see, for example, Fothergill and Gudgin 1982). Such groups may only exist in the classifier's mind, in the sense that their members do not objectively connect to form a coherent group. Intensive research focuses mainly (though not exclusively) on groups whose members may be either similar or different but which actually relate to one another causally; e.g. firms related 'vertically' through linkages or 'horizontally' through competition.

Both types of research are important but they fulfil different functions, the one primarily explanatory, the other primarily descriptive and synoptic. In principle, it ought to be possible for them to be complementary. If a particular kind of mechanism or process has already been identified by intensive research, then provided adequate data

Table 6.1 Intensive and extensive research: a summary

	Intensive	Extensive
Research question	How does a process work in a particular case or small number of cases? What *produces* a certain change? What did the agents actually do?	What are the regularities, common patterns, distinguishing features of a population? How widely are certain characteristics or processes distributed or represented?
Relations	Substantial relations of connection.	Formal relations of similarity.
Type of groups studied	Causal groups.	Taxonomic groups.
Type of account produced	Causal explanation of the production of certain objects or events, though not necessarily a representative one.	Descriptive 'representative' generalizations, lacking in explanatory penetration.
Typical methods	Study of individual agents in their causal contexts, interactive interviews, ethnography. Qualitative analysis.	Large scale survey of population or representative sample, formal questionnaires, standardized interviews. Statistical analysis.
Are the results generalizable?	Actual concrete patterns and contingent relations are unlikely to be 'representative', 'average' or generalizable. *Necessary* relations discovered will exist wherever their relata are present, e.g. causal powers of objects are generalizable to other contexts as they are necessary features of these objects.	Although representative of a whole population, they are unlikely to be generalizable to other populations at different times and places. Problem of ecological fallacy in making inferences about individuals.
Disadvantages	Problem of representativeness.	Lack of explanatory power. Ecological fallacy in making inferences about individuals.

exist it might be possible to use extensive methods to discover its incidence and extent (see Table 6.1). In isolation, extensive research often fails to indicate what processes have produced the patterns it reveals (e.g. *why* the rural/metropolitan contrast in manufacturing employment). Conversely, intensive research does not tell us how widespread or 'representative' are the results produced by the particular process on which it is focused, though that is not its purpose. The possibility of relating intensive and extensive research depends very much upon the homogeneity of the population and the relevance of the categories used to define the characteristics of the sample or population; i.e. whether the properties or variables have clear *causal* significance (e.g. type of industry) or no readily apparent significance (e.g. 'rurality') (cf. Keeble 1980). For example, a popular variable in extensive industrial research is plant size. We have to decide whether this is a meaningful dimension on which to differentiate plants and hence whether any regular associations between this variable and others are causally significant or accidental or whether the variable is a surrogate for something else. In intensive studies it is easier and indeed necessary to get behind the surrogate and find something that is causally relevant to location and performance, e.g. in the case of *firm* size it might be ability to raise capital for major investments.

In the absence of intensive studies one can speculatively examine all manner of formal, quantitative relationships between various characteristics (e.g. firm size, mobility, location of origin and destination, etc.) without getting much closer to the causes of the patterns elicited by the regressions, correlations, etc. If we wonder whether an above-average concentration of R & D in a particular region is an effect of its industrial structure, it is usually possible to do a further extensive analysis (perhaps on a shift-share basis) to consider the 'structural effect'. But whatever the answer the interpretation is *still* likely to be ambiguous[8] precisely because the method is still identifying plants taxonomically rather than causally, seeking out formal quantitative regularities among objects which have no real connection even if they have similar attributes.

In other words, the extensive method lacks explanatory penetration not so much because it is a 'broad-brush' method and insufficiently detailed, but because the relations it discovers are formal ones of similarity, dissimilarity, correlation, etc., rather than substantial, causal relations of connection.[9] For example, we could have used an extensive method for studying modern industry in Wales, and indeed, for the purposes of providing background descriptive information, we found it useful to do so. But no matter how thorough, it is difficult to *explain* much from this kind of analysis. Unless we can identify what specific substantial, causal relations identifiable agents entered into we are liable to be still unclear why plants A–Z each performed as they did. And variations in performance rarely correlate clearly with the usual 'variables' chosen in industrial and regional geography.

But in turning to intensive research one is not necessarily overwhelmed by detail, complexity and differentiation. It is often relatively easy to explain the variations in performance by reference to such things as market shares, achievement of economies of scale, process and product innovations, difference in output per worker, response to excess capacity, changes in licensing arrangements, etc. By looking at firms in contexts which are causally relevant to them and examining what they actually *did*, the logic or structure behind what seemed to be inexplicable patterns in the aggregate data becomes clear. Often intensive sector studies of this nature already exist in industrial economics. It must be admitted however that this kind of analysis is much easier for sectors involving production for mass markets (e.g. domestic electrical appliances) than for complicated, highly differentiated sectors producing mainly custom-built or at least highly specialized commodities, such as defence electronics, because the behaviour of individual firms is less interdependent. However, for the same reasons, intensive analysis is all the more necessary in such cases, because the likelihood of the 'logic of the situation' being evident from extensive studies is smaller.

What inferences can be drawn about individuals in extensive research and about general patterns in intensive research? The former inferences are limited by the problem of the ecological fallacy: combinations of characteristics true of aggregates are not necessarily true of the individuals that comprise them. A converse, unnamed fallacy exists for intensive studies in attempting to draw conclusions about an entire population, although they are not intended to produce results which are 'representative', just as extensive studies are not designed to tell us about individuals. Does this mean then that the findings of intensive studies give us little or no idea of what is happening elsewhere?

In so far as the research deals with the relations between phenomena or characteristics which are contingent (i.e. neither necessary nor impossible), statements about them are unlikely to be generalizable. In so far as intensive research identifies *necessary* relations or properties of objects, then one can say wherever those objects exist then so will those relations or properties. *Actual concrete processes or events are produced through a combination of necessary and contingent relations and so the research findings describing these are unlikely to be generalizable to other contexts.* In our study then, we do not expect the description of concrete events in electrical engineering in South Wales to be generalizable though we expect the mechanisms and structures generating these events to be found elsewhere. Others might therefore combine an understanding of these structures and mechanisms (e.g. the law of value) with information on contingently related conditions specific to different situations in order to explain their concrete patterns.[10]

It should also be noted that although intensive studies are more restricted in the number of individuals they can study, this does not mean that they can only be applied to small scale phenomena like individual households. A multinational firm is also a causal group, but hardly small in scale. So in our case, even though the 'target' of the research is South Wales, it is not necessarily parochial.

One very general implication for policy evaluation of the two types of research design should already be clear. *Because intensive studies allow the identification of causal agents in the particular contexts relevant to them, it provides a better basis than extensive studies for recommending policies which have a 'causal grip' on the agents of change.* Extensive research aids policy analysis by picking out general trends and patterns synoptically. But while contrasts between say industrial change in 'inner-cities' and 'free-standing towns' are striking enough, their revelation does not lead easily to clear policy suggestions precisely because they do not deal sufficiently directly with the agents of change.

Interviews

We are interviewing 'both sides' of industry because we want to learn about their different interests, perceptions and responses. Interviewing only managers reinforces the 'view from above' which already saturates official statistics and most discussions of regional development. At the same time we are limited by problems of feasibility in interviewing workers of different types, not to mention the ethical questions of taking up their time without being able to offer anything directly useful to them in return.

We are wary of seeing managers, unions and especially workers in the stereotypical terms which are all too common in writing on industry. In the case of workers these tend to oscillate between extremes of the recalcitrant, combative workers sold out by the union bosses and the passive, supine workers who fail to be moved by the tireless activists! These stereotypes arise through a failure of the investigator to listen and a tendency to criticize behaviour without understanding the reasons behind it.[11] *However, critical evaluation cannot be dispensed with or tacked on as an optional extra under the heading of 'policy implications' for in assessing the adequacy of various explanations offered by different groups of their activities, we inevitably have to judge which of these are more or less correct.* The reasons for low unionization among women workers is a good example: the explanations given by unions, women workers themselves and management differ. To sit on the fence as regards the evaluation of the conflicting views is also to fail to give a coherent explanation (cf. Sayer 1981). In this context, we are particularly interested in the extent to which unions represent the interest of workers of different types.

Interviewing 'both sides' of industry also necessitates a certain chameleon-like behaviour on our part which might seem to inhibit us in evaluating the role of various groups and individuals. But on methodological and theoretical grounds we feel it is inappropriate to criticize individuals anyway. What managers and others do is very much a function of the pressures and constraints which bear upon them, and *it is not the individuals but the form of social organization which we would expect to criticise*! One can interview managers who have just decided to make hundreds of workers redundant without coming away with the impression that as individuals they are evil. Simply to blame (or praise) individuals *as* individuals, is not so much ethically dubious as explanatorily inadequate and politically naive.

In any interview there is inevitably a significant element of 'impression-management' or 'self-presentation' (cf. Sayer 1981) – and not just on the side of the interviewee. People usually try to answer in a way which puts their actions in the best possible light, though this of course may shade into deceit. Given the conflicts of interest it would again be naive not to be sceptical of information that has not been corroborated by others. In this respect, extensive methods are at a disadvantage: although they can *replicate* the same question for large numbers of different agents they cannot corroborate and enrich the information about a particular event cited by an identifiable individual. For example, a particular type of answer to a question in an extensive industrial survey may be widely replicated, e.g. 'our plant needs more space', but without an intensive study of the multiple determinants of this need as they exist in particular plants one can fail to appreciate how the answers denote different processes underlying what is only superficially a common characteristic.

We chose not to use formal, standardized interviews and questionnaires, which are most favoured in extensive studies. The rationale behind the orthodox method is that by asking each respondent the same questions under formal, controlled (i.e. quasi-experimental) conditions one minimizes observer-induced bias and allows controlled comparisons. But such ideas sacrifice explanatory penetration in the name of 'representativeness' and 'getting a large enough sample'. Extreme standardization which disregards the different types of respondent can in fact make comparisons rather meaningless, because they fail to register the fact that different questions can have a vastly different *significance* for firms of different kinds (and not just according to sector). This not only runs the risk of boring interviewees so that they reveal the minimum, but is also likely to produce results which differentiate firms on criteria which are not necessarily materially relevant to them: i.e. it allows them to be compared taxonomically but not causally. There is also a growing literature on methodology in social science which draws attention to the fact that interviews are *inevitably* interactive. Rather than attempt to minimize interaction (in the hope that observer-induced bias will be reduced) – which only leads to an awkward and often distorted form of (non?-)communication – it is better to *use* the interaction consciously to maximise information flow. This is obviously not a licence to try to influence the interviewee but it does involve changing questions and emphases during the interview in the light of what the subject can talk about. It also ensures a higher response rate! (Cf. Brenner *et al.*, 1978, Harré 1979, Oakley 1981.)

Theory and policy

In capitalist economies firms do not prosper or fail in isolation, in terms of some absolute criterion, but only in terms of how they compare with competitors. (It is not enough merely to be able to produce something that is needed, it must be produced in such a way that adequate profit can be made, in competition with others.) As long as production is organized on this basis, then one firm succeeds (or fails) only where and because others have done worse (or better). There are therefore relations of *interdependence* between firms in different sorts of health. Firms, and hence local economies or regions, can only succeed in becoming 'more competitive' if others become *less* competitive. Only in the usually short-lived cases where markets are expanding extremely rapidly and/or are highly segmented is inter-firm competition in product markets relaxed, and even then it is latent, by virtue of the fact of separately run units of production and freedom of the buyer to choose suppliers. In any case, even where each firm has its own (highly specialized) product market, they still compete for investors' money (e.g. some of the subsidiaries of companies like Racal–Decca, Plessey and Ferranti, involved in producing custom-built defence equipment are certainly competitors for investors' money, even if they each have their own market niches. Indeed, in investment markets they compete with a much wider range of firms than they do in product markets).

In other words, there is an *element* of a zero-sum game in the situation, or to put it more accurately, it is a game in which net gains gradually (and irregularly) accumulate but which necessarily involves losses to some participants, often, as has largely been the case with problem regions, to the same ones repeatedly. The opening of a new factory in place *A* is not necessarily simply a discrete addition to the stock, nor for that matter is a closure necessarily simply a loss and no more. The opening of a 'more competitive' plant at *A* may unknowingly help cause the closure of another plant at *B* which has now been rendered 'uncompetitive', and the closure at *C* may temporarily take the heat off one of its competitors at *D*. Perhaps the clearest contemporary illustrations of this kind of interdependency are *within* the British Steel Corporation, but it also exists of course *between* firms.

Given this interdependency we have a situation in which it is spurious to assume that because some individuals can gain, all individuals can gain simultaneously. In such cases, this assumption that what is possible for the individual is possible for all individuals simultaneously is called 'a fallacy of composition' (Elster 1978, 98). (A more obvious example is the fallacy, crucial to competitive forms of education, that every student can win first prize simultaneously.)

Extensive research designs which search for generalizations about common properties and quantitative changes fail to identify such interdependencies, although they can register their effects 'in the data' unknowingly. As such they are extremely vulnerable to the fallacy. Such procedures tend to support the traditional conclusion that the solution to the regional problem is simply to alter the distribution and quantity of plants, as if their individual fortunes were independent. Superficially, such a conclusion might have seemed plausible in an era (like the post-war boom) when output growth meant employment growth, but since the late 1960s most manufacturing industry has been characterized by jobless growth (and, since 1979, by output decline too). Now at the level of diagnosis of the regional problem, we think all commentators fully recognize that we are experiencing a catastrophic *national* decline in manufacturing employment, yet many of the policy recommendations fail to come to grips with this (e.g. Rothwell 1982[12]). For example, if R & D and science parks generate increases in employment (and on a supra-regional scale we doubt whether they are doing so now) why should they do so more effectively in Warrington than Woking? If the idea is to keep Woking's R & D going and *add* R & D at Warrington then we must reckon with the structural limitations on the amount of such activities capitalist economy (in crisis) can sustain. (There are apparently now between 30 and 40 British towns planning science parks! (*The Guardian*, 5 January 1983).) And it also has to be shown whether the output of R & D improves or worsens the employment situation in the economy as a whole. If the policy does not make any difference to the total, it only relieves the regional problem in the sense of possibly equalizing the misery a little more. In some cases (e.g. local initiatives to attract industry) the beggar-thy-neighbour character of such policies is widely recognized, but it has yet to be detected widely in its more sophisticated variants.

These are just a few of the diverse assortment of suggestions which have been made in response to regional and industrial problems. We now want to argue that despite the diversity of responses, they can be shown to fall into three types or levels by means of a simple model (derived from Elster 1978, 135).

REACTIVE AND/OR BASED ON MISUNDERSTANDING (LEVEL 1)

The nature of the situation or 'game' is either (a) not understood, or (b) misunderstood. In the case of (a) many workers lack the time and information to do more than *react* to problems as and when they hit them. In (b) we find many academics, politicians and also some managerialist positions which misread the situation as an unqualified positive sum game in which any problems are discrete, residual and potentially soluble within the rules of the game, i.e. they fall for the fallacy of composition, and not always unknowingly.

TACTICAL/PLAYING THE SYSTEM (LEVEL 2)

The nature of the game and the problems it necessarily generates are roughly understood, but actors and commentators use this understanding to try to play the system to their own individual advantage, knowing full well that not everyone can gain simultaneously and indeed a substantial proportion will lose.

COLLECTIVE (LEVEL 3)

Actors and commentators understand the nature of the game but refuse either to play the system individualistically or to reason fallaciously that all can gain simultaneously. Instead they realize the possibility of mutual gain by replacing the competitive structure by a collectively planned one, i.e. by changing the game.

Actors' and groups' views also tend to differ according to their position in the game. Management and its allies respond at levels 1 and 2, often reasoning by the fallacy of composition; 3 is obviously not in their interests, though sometimes limited co-operation is suggested in order to combine forces against others, usually foreigners (e.g. NEDO, MITI . . .?). From the point of view of a British firm, whether imports from cheap labour countries are a good or bad thing depends on whether it is in on the act. Strong multinational firms tend to resist calls for protectionism, weak national firms tend to support them. Attitudes towards *prospective* inward investors which threaten domestic industry differ radically if they succeed in becoming part of domestic industry. For example, when Hitachi wanted to build a TV plant in Britain, NEDO conspired with 'British' (some of them foreign-owned) firms (and unions) to keep them out, but once they got here (by collaborating with one of the firms that formerly tried to keep them out!), NEDO switched to supporting them. In sectors of industry, experiencing jobless growth, capital argues against attaching employment conditions to the allocations of development grants such as RDGs, but (depending on the labour needs of the particular firm) objects to them less where output and employment growth are more compatible. In other words diagnoses of problems and policy recommendations may differ markedly even at the same level (in this case level 2) yet can be shown to derive from different but nevertheless interdependent positions in the same game.

Too many academic analyses and prescriptives regarding the regional problem fall into categories 1b and 2. This is not necessarily because their authors sympathize politically with the interests of capital – in fact many do not. But then explanations and prescriptions are not simply a function of their authors' political sympathies; they are also influenced by theory and method, and if this obscures causal structure then general political sympathies and actual policy prescriptions may diverge.

MANAGEMENT VIEWS

Consider the following types of proposal which might be proposed by managers or their (often unaware) supporters:

(a) the solution is to increase productivity, to match or improve on productivity and product-quality standards of the Japanese, to make the regions' industry more competitive;
(b) if workers are too militant in defence of their position, potential foreign inward investors will be scared off;
(c) workers should moderate wage demands (i.e. concede wage cuts in real terms) in order to make British industry more competitive (price themselves into work/poverty?);
(d) all regions should strive for more R & D and early product cycle production so as to get the benefits of both higher incomes in this field and of modernized production through the application of new technology;
(e) regions should increase local, intra-regional linkages between firms.

All these proposals and many similar ones rest upon fallacies of composition.[13] Yet, *from level 2 management standpoints*, they make sense, and indeed it is possible to distinguish better from weaker policies at this level, e.g. massive long-term investment in R & D and technological change is more likely to bring success for individual capital than relocation to enterprise zones. It is also clear from these proposals that the interests of capital and labour are at odds too. As a result they provide (individualistic not generalizable) ways of producing 'development' *in* a region but not development *of* a region, indeed some of them – (b) and (c) – encourage the former at the expense of the latter. These proposals also appeal to the same mechanisms which have produced uneven development to remedy its effects.

UNION VIEWS

Union responses are usually pitched at levels 1 and 2 with some efforts (though often largely at the level of rhetoric) in the direction of 3. Often they lack the information and resources necessary to be able to play the system individualistically and are forced into a reactive role. Often they are tempted to criticize capital on its own terms at level 2, e.g. 'why did the firm not invest and diversify its product range?', and there are many cases where the specific reasons why a *particular* plant failed can be attributed to such management errors.

However, this does not explain why, in many situations, elimination of *some* plant or other is a systemic or structural effect, e.g. ITT workers at Hastings might be tempted to blame the closure of the television factory on the failure of their management to compete successfully against other firms, especially the new Japanese plants established in Britain. But while they may be correct in their diagnosis of why *this* plant rather than another failed, they do not come to terms with the systemic or structural reasons why there had to be any closures at all, i.e. overproduction of TV producing capacity as a result of unplanned private production, exacerbated by pursuit of rapidly rising optimal scales of production, etc.

In the nature of the 'game', despite the contradictory interests of capital and labour and the oppositions between competing capitals, it is possible for some capital in some conjunctures to be significantly more favourable than others, e.g. employment in a leading multinational 'big league' firm may be better paid and more secure than in a national 'little league' firm. This unevenness again invites arguments based on fallacies of composition, and tempts unions to accept weak agreements (even no-strike clauses) in order to attract what are perceived to be relatively secure employers, although obviously the general weakness of labour in the recession has a lot to do with this. Some of the more right wing unions, particularly the one of most relevance to our research, the EEPTU, have largely echoed management's level 2 type arguments about policy. Radical analyses often underestimate the importance and ideological significance of this form of commonality of interest. Obviously the dilemma, that opposition to possible future closures, speed-ups or redundancy may make them more rather than less likely, is real enough. It is in the nature of the very structure of the game that such strategies are highly risky. But as long as we remain at level 2 the dilemma will remain too.

There is an element of truth in the argument that many union practices, such as free collective bargaining, may be informed by a fallacy of composition, and that in being uncoordinated are not unlike the free-for-all of the competition between firms. The similarity allows union level 2 responses to play up the ironic similarity ('firms are expected to get the best deal for their shareholders, unions the best deal for their members – they try to maximize their earnings and so do we, and so we are really in the same game'). Equally it is possible for apologists of capital to hypocritically point out the fallacy of composition ('What is possible for the miners cannot be possible for all workers simultaneously'). Just as labour can reply that free collective bargaining is not a zero-sum game because it can simultaneously benefit all workers indirectly by boosting aggregate demand, capital can reply to the charge that it is involved in a zero-sum game by pointing to the increase in wealth that capital accumulation brings. Both arguments have both an element of truth and a *non sequitur* (increased aggregate demand may largely suck in imports, stimulate inflation and any benefits certainly will not be equitably distributed, while the benefits of further capital accumulation need not accrue to workers in general or even to those most responsible for it). Such points are very much the stock-in-trade of political debate on the economy. Again within the terms of level 2 responses more or less rigorous and coherent versions of these arguments can be distinguished. But what is important to recognize is how the alternatives are 'trapped' within the logic or structure of a capitalist organization of production. In particular they are trapped within the contradiction in which it is in the interests of any one firm for all other firms to pay higher wages, as it increases demand, but not to pay its own workers more, as it cuts into profits. Historically unions have developed within and had to accommodate to the structure which dictates level 2 strategies.

But unions do not directly represent the interests of the unemployed and may harm them where they allow jobs to be sold in order to protect those individuals who want to keep their jobs. We do not underestimate the dilemmas of this situation (having been personally involved in them!) but once again it makes no sense to pretend that an individual solution ('Let's allow voluntary redundancy so that the rest of us can keep our jobs') is a general, collective solution. In addition unions do not always represent those workers who *are* their members. Here we are not thinking so much of the classic 'sell-outs' of the rank and file by the union bosses, but rather the way in which unions have institutionalized assumptions of a particular kind of worker – the male breadwinner with 'unemployed' wife and children to support – who is no longer representative (if he ever was). Particularly in South Wales and in electrical engineering there has been a dramatic increase in female employment. Even though women may often be the only breadwinner supporting a family, their position and constraints are different from those of male breadwinners, in that they are still expected to do a 'second-shift' of unpaid work in the home. Their evaluations of the relative benefits of more pay, workplace facilities (e.g. creche, busing), overtime or shorter hours are therefore very different from those of men. (To report men's lack of interest in creches or 'shorter hours so as to have more time with the family' is not to support it.) The particular kind and extent of domestic commitments can also make a significant difference to the usefulness of the worker to capital, and we strongly suspect that personnel managers are becoming increasingly aware of the advantages of vetting applicants for jobs to check that their domestic commitments do not interfere with their performance as workers. Indeed, as competition intensifies in the recession, this has become a necessity for firms.

Both roles – that of the traditional male breadwinner and the female double-shift workers – are interdependent. Jointly they reproduce and are reproduced by sexist assumptions in the home and the firm. We are *not*, therefore, suggesting that unions accommodate to the needs of women workers *simply as those needs are generated by this division of labour* (e.g. by negotiating for shorter hours and creches for women parents only) for to do so would perpetuate the inequality which derives from the gendered division of labour, paid and unpaid. Rather they should push for changes in paid work which allow equality in the home and not merely in paid work. Equality in one sphere is not possible without equality in the other; union (and worker) connivance at gender inequality in the firm serves to reinforce it at home.

Many industrial studies have failed to notice either these conflicts of interests between unions and workers or the reasons for the increasing numbers of women workers in this kind of industry. In our interviews and questionnaires we are paying special attention to these issues.

From the point of view of industrial and regional geography such a focus may seem unusual, but this strangeness bears witness to the lack of questioning of the (male) 'view from above' in orthodox studies and their neglect of the difference between development *in* regions (from the standpoint of capital) and development *of* regions (from the standpoint of different kinds of worker).

Earlier we summarized level 3 responses all too briefly and glibly.

Not surprisingly though, given that it involves fundamental structural change they are extremely difficult to implement. Often they cannot be constructed simply as a continuation of the kinds of policy developed at level 2 but involve a break with, and in some cases virtually a reversal of, level 2 policies.

The most fundamental aspect of level 3 responses is the need (i) to develop a planned system of production and (ii) to ensure that plans are made in accordance with needs. There is much teetering on the brink between levels 2 and 3, so that (i) is sometimes accepted without (ii) in which case plans are made primarily to make British capitalism more efficient and profitable – a strategy that would only reproduce regional inequality. Without (ii) the state merely enlarges its say in decisions made primarily according to capitalist criteria. The blatant contradictions between state policy *vis-à-vis* steel and coal, and the (expressed) policy *vis-à-vis* the Development Areas in which much of these industries are located demonstrates only too clearly the inadequacy of (i) without (ii). If level 3 proposals for a planned national economy are coupled with proposals to make British capital more competitive then attempts to produce development *of* the regions will inevitably come a poor second and will only be successful where regional interests just happen to coincide with the needs of capital.

Such a situation would not be very different from the present one. Now, over 90 per cent of regional aid takes the form of Regional Development Grants having no strings attached as regards employment. They may not lead to employment increases and indeed can even induce losses through *in situ* automation and/or negative effects on competitors. At the same time regional policy is still presented as a means to the development *of* the region. As was noted in the Public Accounts Committee:

> the main policy instruments . . . have been designed to operate on investment rather than directly on employment . . . [and] increases in employment do not proceed *pari passu* with new investment. (Committee of Public Accounts, HC 206)

So while we think it essential that regional planning (on a nationwide basis) be integrated into a national economic strategy, the success of such a strategy would depend on what goals and criteria would be used in the planning and whether they could be enforced. In the former respect, Labour's recently published *Alternative Regional Strategy* discussion paper is very forward looking, but it is equivocal about the latter. Similar kinds of equivocation are common in academic analyses too, and inappropriate research methods can contribute to the uncertainty by concealing the causes and agents of uneven development.

Some radical commentaries also oscillate between level 2 and 3 responses. For example, the Community Development Programme's *Costs of Industrial Change* report and the Counter Information Services reports on industry tend both to criticize particular capitals (often personalized as attacks on particular capitalists in the CIS reports) for being inefficient, *and* to criticize, at least implicitly, the *system* which gives such judgements their rationale. Although the former stance is very tempting it is hardly consistent.

In so far as our study involves a comparison of the performance of firms in the sector in South Wales, it is easy to slip into a level 2 type of criticism. Some of the literature on the sector which we have found most useful takes such a stance, and this, coupled with the glaring differences between some of the highly efficient Japanese firms and traditional British (and American) firms in Wales, makes it tempting to echo their arguments (e.g. 'the only hope is for other firms in the region to emulate the Japanese in efficiency and quality'). Yet such proposals can only offer the prospect of redistributing, not solving, the problems of uneven development.

Given the radical differences between levels 2 and 3 some of the often embryonic level 3 proposals are bound to seem Utopian and unrealistic from level 2, but we believe that instead of dismissing them the constraints which make them seem so must be understood. *If workers' plans for production geared to need are not feasible we must ask what must be changed to make them feasible?*

Attempts to find alternative ways of defining and calculating economic rationality need to be criticized constructively rather than dismissed. There are a number of variations on the theme of 'social costs assessments'. There was the Ward and Rowthorn assessment of the social costs of steel closures in 1979 which found that forcing BSC into profit would cost considerably more nationally than it saved within the company (Ward and Rowthorn 1979), and this phenomenon is surely not unusual. There is the 'social accounting approach' of the West Midlands Enterprise Board with its criterion for investment of return to the (social) *community* as a whole rather than to any single (private) *investor*. Both these approaches are steps in the right direction, we feel, but both must be recognized as only making a partial achievement of forms of economic calculation which respond to need rather than profit and purchasing power. Likewise we must realize how common it is for agencies which were originally set up ostensibly to encourage investment to satisfy (regional) needs, to drift into operating according to criteria not unlike those of a merchant bank, and hence for goals compatible with development *of* regions to change into goals favouring development *in* regions.

Conclusions

If research is to produce politically relevant information it must pose its questions and choose its methods in a way which allows answers of the right form to be produced. Discovering the extent of a problem or process is different from explaining its origins. We must decide which we want to know and design the research accordingly. We believe that the theoretical and methodological perspectives embodied in our research afford us a greater purchase on the agents and processes of regional development and the conflicts between them, than the relatively bloodless categories of aggregate statistics. Nevertheless, extensive methods have provided useful identifications of the changing patterns to be explained. Far from being a wholly academic matter, the question of method is of crucial political importance in generating information that can be socially useful.

Notes

1 We are grateful to the ESRC for funding this research.
2 SIC=Standard Industrial Classification. SIC Order IX is electrical engineering.
3 MLH=Minimum List Heading – sub-sectors within each SIC order.
4 This distinction is borrowed from Harré (1979, 132).
5 The debate between Sayer and Keeble (Sayer 1982b) basically amounts to an argument about the possibility of evaluating intensive research in the same way as extensive research. 'Explaining manufacturing shift: a reply to

Keeble', *Environment and Planning A*, *14*, 119–25, with a response by Keeble. See also Keeble (1980).

6 Ethnographic approaches are generally intensive too.

7 This term is adapted from Harré (1979).

8 See Massey and Meegan (1982, 191) for a discussion of these ambiguities.

9 In effect, this is another way of contrasting explanation by generalization with explanation by abstraction. Cf. Sayer (1982a).

10 Actually it is not uncommon for the unusual case to reveal more about general processes than 'normal' cases. (For example in urban geography – the company town; in psychology – identical twins reared apart.)

11 That is 'that kind of criticism which knows how to judge and condemn the present, but not how to comprehend it'. Marx (1976, 505).

12 Rothwell's position is all the more extraordinary because having been prominent in drawing attention to the phenomenon of jobless growth (e.g. Rothwell and Zegveld 1979), his recent prescriptions appear to take no heed of it (Rothwell 1982).

13 The arguments about these kinds of proposals are discussed more systematically in Sayer (1983).

7 Postscript: Doing research

We may not have been able to agree on the answers but what we hope we have been able to do is to make clear some of the questions that must be addressed when setting out on a research project. The approaches that have been elaborated here are not hermetically sealed from each other – we proved that by at least being able to have a debate across their boundaries. But they are coherent and distinctive packages which run the gamut from the design of the research question, the choice of methodology, the selection of data to the very categories and concepts which can validly be used. It is not therefore possible to be unthinkingly eclectic. Intensive research *really is* completely different from extensive research, for instance, and it is essential to know which approach you intend to use. They feed into each other but effort is needed to make them genuinely complementary.

Although the contributors to this collection would suggest different solutions to the above problems, we are all agreed on the importance of getting to grips with, and answering, the following issues at an early stage of research design:

- what kind of an explanation are you looking for? – where do you expect it to be located? – how is it structured?
- are you looking for patterns or processes? – what does that mean for your research design?
- are the categories and concepts you are looking for consistent with these aims?
- in what way do you want your data to be comprehensive – in coverage of areas or industries? – in the detail of investigation of individual corporations perhaps? – in the range of actors you investigate?
- if you see the necessity for different levels of analysis, what exactly do you want from each level? – how are the levels to be linked ? – and (to avoid the age-old problem of feeling one ought to be studying the cosmos) how is each to be delimited?
- if there are different actors involved in the issues you are studying how are you to structure the interaction between them? – how do you propose to analyse any conflicts of interest or information which may exist between them?
- how and in what way do you want your conclusions to be generalizable?

These are just a few issues, but all of them are essential to answering what is probably the most difficult research problem of all – what exactly is the research question you are trying to answer?

Reflections on policy

You may or may not wish your research to come up with conclusions for action. The different contributions to this collection have ranged over a wide spectrum. The following are some of the broad aspects of this question which we all felt bound to tackle.

Spatial scale Capitalist production is organized and operates at international, national and local level and probably all the contributors here would agree that there is a need for policy at all of these levels, albeit with differing emphases between them. It is indeed somewhat ironic that while there has been increasing recognition over the past decade or so that the causes of local economic decline are to be found at international level, there has been an upsurge in interest in policy formulation and operation at the local level. And this interest spans the political spectrum. Thus the Right argues the case for local small business centres, the Centre for local profit-making co-operatives and the Left for local 'alternative production'. But it is perhaps especially ironic for the Left, on the one hand to have expanded the explanation to the level of international capitalism and then to have come up on the other with local 'bottom-up' community planning. Can such a local strategy be effective if the causes of local economic decline, and the resources needed to deal with it in a comprehensive manner, are to be found at broader spatial scales? One reply to this is the point that local strategies can be formulated which deliberately operate within the interstices of the local market economy through non-profit-making co-operative ventures and local community plans. It is certainly the case that local authorities and local trade unions have joined forces and linked locations, to fight the production reorganization strategies of multinational companies. There is scope for a local response. But another reply is to raise the question of what anyway is meant by 'effectiveness' in the case of policies for local alternative production? Is it simply job creation or is it more than this? Local initiatives of this kind are argued to have an important political role to play – in that they provide a vehicle for building alliances at local level and the confidence that production can be restructured towards meeting community needs, that jobs and technology can be designed to match the aspirations of the people who actually perform them. Similarly it is also argued that it is only by building up policy from the bottom that a truly democratic approach to restructuring at wider spatial scales can be achieved.

Whether policy should be focused on areas or on production The answer to this is, of course, crucially dependent on one's analysis of the causes of regional uneven development. Are these to be found in the characteristics of the regions themselves, or are they to be found in production? In this collection both views have been put forward. It has been argued that policy has to operate within a system of social relations that is given – capitalist production. In this view, regional inequalities can only be addressed by policies which either seek to change areas in such a way as to make them match more adequately the requirements of this system or which recognize that some areas will always lag behind others and need support on social grounds. Such considerations have indeed set the whole tenor of incentives-based regional policy in this country. Area-based policies are certainly politically much easier to operate than are policies involving direct intervention in the control and spatial organization of production. But, alternatively, others have argued that if the *causes* of regional inequalities are to be found in the system of capitalist production then there is no alternative but to intervene in this system – whatever the political difficulties of such a policy course.

Policies towards the corporate sector The next question, then, might be 'how to intervene?' Given that the capitalist economy and its geography are dominated by the activities of large, multi-plant, multinational corporations, what can and should policy be towards these? If it is accepted that policy should go beyond that of simply accommodating to their locational and production strategies and actually attempt directly to influence their behaviour, the question, then, is often seen as a purely tactical one. How can sufficient countervailing power and leverage be achieved? Collaboration between national governments? International combines of trade unions? It could be argued, however, that the question of intervention is a more fundamental one than this. Corporations are seen, in this scenario, as the agents of change. Indeed this is precisely the reason why policies are directed towards them – they are 'to blame'. But blamed for what? One position is to assume that, because of what we have called the behavioural element in their operation, they are somehow acting irrationally and that policy is required simply to make them act in a more rational way. A more common argument is to accept that they are behaving perfectly rationally on their own terms but that policy is still required to make them change their behaviour. The problem then, of course, is what sense is there in making capitalist enterprises operate in a non-capitalist manner? If these companies cannot compete internationally will they, or the parts affected by policy at least, not go under eventually? Does the problem lie not with the 'agents' as such but with the structure of which they form part, and the system in which they operate?

How far should policy go in attempting to change the relations of production? Some argue that capitalism is here to stay. But, as far as policy is concerned, this is more than a simple argument that the country will remain fundamentally a mixed economy. It tends also to rule out the possibility of ameliorative reform (rather than revolution) based on either altering capitalist social relations within specific parts of the economy or on accepting areas of non-profit-oriented production (via social audits, etc.). Of the contributors here who do recommend policies which challenge capitalist relations of production none expect any immediate overturning of them across society as a whole. Yet they still argue that such policy initiatives are necessary. The question they are faced with is, can islands of socialism survive in a sea of market forces? How can alternative forms of production be established and pushed forward?

To whom should policy recommendations be addressed? A glance at the economic geography literature will reveal the extent to which policy findings are aimed at a somewhat prescribed list of agents, namely central and local government and companies. The increasing concern with the geography of job loss has meant that labour features more prominently in the analysis, but often it is assumed to be unorganized and/or on the receiving end of policies formulated by other agents. Indeed policies aimed at labour from the Right are often formulated precisely to deter any interference by labour – through militancy, wage demands, or drawing up their own plans – in the strategies of these other agents. Implicit in this argument is the belief that the controllers of production 'provide jobs' and that there is no inherent conflict of interest between capital and labour. This view is challenged on the Left and a much more active role in policy-making is given to organized labour, ranging from shop-floor policies towards existing jobs to the formulation of alternative production strategies. The question then, of course, becomes – how realistic such policies are in the immediate term, and how they would fit into a wider policy context.

Who and what is policy for? A question which nicely encapsulates all the issues we have briefly raised above! Is policy about encouraging development *in* a region or development *of* a region? Is it about making areas suitable for the further development there of capitalist production and its attendant social relations (development *in* a region) or should it be designed on the basis of the needs of the people living there (development *of* a region)? Are these two alternatives compatible, mutually exclusive, or only accidentally coincidental?

Bibliography

Aaronovitch, S. (1981) *The Road from Thatcherism: The Alternative Economic Strategy*, London, Lawrence & Wishart.

Aaronovitch, S. and Sawyer, M. (1975) *Big Business: Theoretical and Empirical Aspects of Concentration and Mergers in the United Kingdom*, London, Macmillan.

Aaronovitch, S., Smith, R., Gardner, J. and Moore, R. (1981) *The Political Economy of British Capitalism*, Maidenhead, McGraw-Hill.

Birch, D.L. (1979) *The Job Generation Process*, Cambridge, Mass., MIT.

Blackaby, F. (ed.) (1978) *De-Industrialization*, London, Heinemann.

Bluestone, B. and Harrison, B. (1980) 'Why corporations close profitable plants', *Working Paper for a New Society* 7, 15–23.

Blunkett, D. and Green, G. (1983) 'Building from the bottom: the Sheffield experience', *Fabian Tract 491*, London, Fabian Society.

Brenner, M., Marsh, P. and Brenner, M. (1978) *The Social Context of Method*, London, Croom Helm.

Broadbent, T.A. and Meegan, R. (1982) 'New technology and older industrial regions in the UK', London, CES (mimeo).

Brown, A.J. (1972) *The Framework of Regional Economics in the UK*, Cambridge, Cambridge University Press.

Brown, W. (ed.) (1981) *The Changing Contours of British Industrial Relations. A Survey of Manufacturing Industry*, Oxford, Blackwell.

Campbell, M. (1981) *Capitalism in the UK*, London, Croom Helm.

CEPR (1981) 'Economic Policy in the UK', *Cambridge Economic Policy Review* 7 (1) April.

Champion, A.G., Gillespie, A.E. and Owen, D.W. (1982) 'Population and the labour market with special reference to growth areas in the UK', paper presented to British Society for Population Studies Conference, Durham.

Cheshire County Council (1982) *Chemicals North-West.*

Cooke, P., Morgan, K. and Jackson, D. (1984) 'New technology and regional development in austerity Britain: the case of the semi-conductor industry', *Regional Studies*, 18 (4), 277–89.

Courtaulds (1982/3) *Annual Report to Employees*, London.

Cross, M. (1981) *New Firm Formation and Regional Development*, Farnborough, Gower.

Damesick, P. (1982) 'Regional problems and policy in Britain: a case for reappraisal', *Built Environment* 7 (2).

Danson, M.W., Lever, W.F. and Malcolm, J.F. (1980) 'The inner-city employment problem in Great Britain, 1952–76: a shift–share approach', *Urban Studies 17*, 193–210.

Dicken, P. (1983) 'Overseas investment by UK manufacturing firms: some trends and issues', *NWIRU Working Paper 12*, School of Geography, University of Manchester.

Dicken, P. and Lloyd, P.E. (1978) 'Inner metropolitan change, enterprise structures and policy issues', *Regional Studies 12*, 401–12.

Duncan, M. (1982) 'The information technology industry in 1981', *Capital and Class 17*, Summer 1982, 78–113.

Dunford, M. (1977) 'The restructuring of industrial space', *International Journal of Urban and Regional Research 1*, 510–20.

Dunford, M., Geddes, M. and Perrons, D. (1981) 'Regional policy and the crisis in the UK: a long-run perspective', *International Journal of Urban and Regional Research 5* (3), 377–410.

Dunnett, P.J. (1980) *The Decline of the British Motor Industry*, London, Croom Helm.

Dunning, J.H. and Pearce, R.D. (1981) *The World's Largest Industrial Enterprises*, Farnborough, Gower.

Elster, J. (1978) *Logic and Society*, Chichester, Wiley.

Erickson, R.A. (1980) 'Corporate organization and manufacturing branch plant closures in non-metropolitan areas', *Regional Studies 14* (6), 491–501.

Evans, A. and Eversley, D. (eds) (1980) *The Inner City*, London, Heinemann.

Fothergill, S. and Gudgin, G. (1979) 'Regional employment change: a subregional explanation', *Progress in Planning 12*, 155–219.

Fothergill, S. and Gudgin, G. (1982) *Unequal Growth*, London, Heinemann.

Fothergill, S., Kitson, M. and Monk, S. (1985) 'Urban industrial change', *Department of the Environment, Inner Cities Research Series*, London, HMSO.

Fröbel, F., Heinrichs, J. and Kreye, O. (1980) *The New International Division of Labour*, Cambridge, Cambridge University Press.

Fryer, R.H. (1973) 'Redundancy, values and public policy', *Industrial Relations Journal 4* (2).

Glyn, A. and Harrison, J. (1980) *The British Economic Disaster*, London, Pluto Press.

Glyn, A. and Sutcliffe, J. (1972) *British Capitalism, Workers and the Profits Squeeze*, Harmondsworth, Penguin.

Green, K., Coombs, R. and Holroyd, K. (1980) *The Effects of Micro-electronic Technologies on Employment Prospects: a Case Study of Tameside*, Farnborough, Gower.

Gudgin, G. (1978) *Industrial Location Processes and Regional Employment Growth*, Farnborough, Saxon House.

Hannah, L. and Kay, J.A. (1977) *Concentration in Modern Industry*, London, Macmillan.

Harré, R. (1979) *Social Being*, Oxford, Blackwell.

Hayter, R. and Watts, H.D. (1983) 'The geography of enterprise: a re-appraisal', *Progress in Human Geography*.

Healey, M.J. (1981) 'Locational adjustments and the characteristics of manufacturing plants', *Transactions of the Institute of British Geographers 6* (4), 394–412.

Healey, M.J. (1982) 'Plant closures in multi-plant enterprises – the case of a declining industrial sector', *Regional Studies 16* (1), 37–51.

Henderson, R.A. (1979) 'An analysis of closures amongst Scottish manufacturing plants', *Economics and Statistics Unit Discussion Paper 3*, Scottish Economic Planning Department, Edinburgh.

HMSO (1983a) *Railway Finances, Report of the Committee chaired by Sir David Serpell*, London.

HMSO (1983b) *Regional Industrial Development*, Cmnd 9111, London.

Hodgson, G. (1982) 'Theoretical and policy implications of variable productivity', *Cambridge Journal of Economics 6*, 213–26.

Holland, S. (1975) *The Socialist Challenge*, London, Quartet.

Holland, S. (1976) *Capital Versus the Regions*, London, Macmillan.

International Federation of Chemical, Energy and General Workers' Unions (1982) *The World Petroleum Refining Industry*, Geneva.

Jensen-Butler, C. (1982) 'Capital accumulation and regional development: the case of Denmark', *Environment and Planning A, 14*, 1307–40.

Johnson, P.S. and Cathcart, D. (1979) 'The founders of new manufacturing firms: a note on the size of incubator plants', *Journal of Industrial Economics 28*, 219–24.

Jones, D.T. (1976) 'Output, employment and labour productivity in Europe since 1955', *National Institute Economic Review* 77, August.

Keeble, D. (1976) *Industrial Location and Planning in the United Kingdom*, London, Methuen.

Keeble, D. (1980) 'Industrial decline, regional policy and the urban–rural manufacturing shift in the UK, *Environment and Planning A, 12*, 945–62.

Kennett, S. (1982) 'Migration between British local labour markets and some speculation on policy options for influencing population distributions', paper presented to British Society for Population Studies Conference, Durham.

Labour Research Department (1984) *Labour Research 73* (1), January.

Lane, T. (1982) 'Dunlop and the world tyre industry', Department of Sociology, University of Liverpool (mimeo).

Law, C.M. (1982) 'The geography of industrial rationalization – the British motor car assembly industry, 1972–82, *Discussion Paper 22*, Department of Geography, University of Salford.

Leigh, R. and North, D.J. (1978) 'Regional aspects of acquisition activity in British manufacturing industry', *Regional Studies 12*, 227–46.

Leontiades, J.A. (1974) 'Plant location – the international perspectives', *Long Range Planning* 7, 10–14.

Lipsey, R.G. (1971) *An Introduction to Positive Economics* (3rd edn), London, Weidenfeld & Nicolson.

Lloyd, P.E. and Cawdery, J. (1982) 'The large manufacturing firm in Liverpool: corporate sector influences on the development of indigenous enterprise', research report submitted to Liverpool Economic Development Corporation.

Lloyd, P.E. and Dicken, P. (1979) 'New firms, small firms and job generation: the experience of Manchester and Merseyside, 1966–1975', *NWIRU Working Paper 9*, School of Geography, University of Manchester.

Lloyd, P.E. and Dicken, P. (1982) 'Local manufacturing firms in the older urban environment: perspectives for Greater Manchester and Merseyside', *Department of the Environment, Inner Cities Research Programme 6*, London, HMSO.

Lloyd, P.E. and Mason, C.M. (1979) 'Industrial movement in North-West England', *Environment and Planning A, 11*, 1367–85.

Lloyd, P.E. and Reeve, D.E. (1982) 'North-West England, 1971–1977: a study in industrial decline and economic restructuring', *Regional Studies 16* (5), 345–60.

Lowry, A.T. (1982) *Pilkingtons: Reflections on an Uncertain Future*, Multinational Business, Economist Intelligence Unit, London, 18–30.

McCrone, G. (1969) *Regional Policy in Britain*, London, Allen & Unwin.

Mackay, D.I. and Reid, G.L. (1972) 'Redundancy, unemployment and manpower policy', *Economic Journal 82*, 1256–72.

Martin, R.L. (1982) 'Job loss and the regional incidence of redundancies in the current recession', *Cambridge Journal of Economics 16* (4), 375–95.

Marx, K. (1976) *Capital*, vol. I, Harmondsworth, Penguin.

Massey, D. (1979) 'In what sense a regional problem?', *Regional Studies 13*, 233–44.

Massey, D. and Meegan, R.A. (1978) 'Industrial restructuring versus the cities', *Urban Studies 15*, 273–88.

Massey, D. and Meegan, R.A. (1979) 'Labour productivity and regional employment change', *Area 11* (2), 137–45.

Massey, D. and Meegan, R. (1982) *The Anatomy of Job Loss: The How, Why and Where of Employment Decline*, London, Methuen.

Minns, R. (1982) *Take over the City: the Case for Public Ownership of Financial Institutions*, London, Pluto Press.

Moore, C.W. (1973) 'Industrial development linkage paths', *Tijd. voor Econ. Soc. Geogr. 64*, 93–101.

Oakley, A. (1981) 'Interviewing women', in Roberts, H. (ed.) *Doing Feminist Research*, London, Routledge & Kegan Paul.

O'Farrell, P.N. (1976) 'An analysis of industrial closures: Irish experience 1960–73', *Regional Studies 10*, 433–48.

Palloix, C. (1975) *L'Internationalisation du Capital: Elements Critiques*, Paris, Francois Maspero.

Panic, M. and Joyce, P.L. (1980) 'UK Manufacturing Industry: International Integration and Trade Performance', *Bank of England Quarterly Bulletin*, March 1980.

Peck, F. and Townsend, A. (1984) 'Contrasting experience of recession and spatial restructuring: British Shipbuilders, Plessey and Metal Box', *Regional Studies 18*, 4.

Prais, S.J. (1976) *The Evolution of Giant Firms in Great Britain*, Cambridge, Cambridge University Press.

Prais, S.J. (1981) *Productivity and Industrial Structure*, Cambridge, Cambridge University Press.

Regional Studies Association (1983) *Report of an Inquiry into Regional Problems in the United Kingdom*, Norwich, Geo Books.

Rothwell, R. (1982) 'The role of technology in industrial change: implications for regional policy', *Regional Studies 16* (6), 361–9.

Rothwell, R. and Zegveld, W. (1979) *Technical Change and Employment*, London, Frances Pinter.

Samuelson, P.A. (1967) *Economics*, London and New York, McGraw-Hill.

Sayer, A. (1981) 'Defensible values in geography: can values be science-free?' in Herbert, D.J. and Johnston, R.J. (eds) *Geography and the Urban Environment*, vol. IV, Chichester, Wiley.

Sayer, A. (1982a) 'Explanation in human geography', *Progress in Human Geography (V) 6* (1), 68–88.

Sayer, A. (1982b) 'Explaining manufacturing shift: a reply to Keeble', *Environment and Planning A, 14*, 119–25.

Sayer, A. (1983) 'Theoretical problems in the analysis of technological change and regional development', in Hamilton, F.E.L. and Linge, G.J.R. (eds) *Spatial Analysis Industry and the Industrial Environment*, vol. 3, Chichester, Wiley.

Short, J. (1981) 'Defence spending in the UK regions', *Regional Studies 15* (2), 101–10.

Shutt, J. and Lloyd, P.E. (1985) 'Industrial restructuring in the commercial vehicles industry', *NWIRU Working Paper 17*, forthcoming.

Singh, A. (1977) 'UK industry and the world economy – a case of de-industrialization', *Cambridge Journal of Economics*, June.

Smith, I. (1979) 'The effect of external takeovers on manufacturing employment change in the Northern Region between 1963 and 1973, *Regional Studies 13*, 421–37.

Smith, I.J. (1982) 'Implications of corporate restructuring in the telecommunications and electronic computer industries during the 1970s', in Goddard, J.B., Gillespie, A.E. and Thwaites, A.T. (eds) *Information Technology and the Less-favoured Regions of Europe*, CURDS, University of Newcastle-upon-Tyne.

Sunday Times (1982) 'Ford fights to close the Saarlouis gap', *Sunday Times Business News*, 20 March.

Taylor, M. and Thrift, N. (1981) 'British capital overseas: direct investment and corporate development in Australia', *Regional Studies 15*, 183–212.

Taylor, M. and Thrift, N. (1982) *The Geography of Multinationals*, London, Croom Helm.

Times 1000, The (1981–2) London, Times Books.

Townroe, P.M. (1975) 'Branch plants and regional development', *Town Planning Review 46* (1), 47–62.

Townsend, A. (1980a) 'Recession and the regions in Great Britain, 1976–1980; analyses of redundancy data', *Environment and Planning A, 14* (10), 1389–404.

Townsend, A.R. (1980b) 'Unemployment and the new government's "regional" aid', *Area 12* (1), 9–18.

Townsend, A.R. (1980c) 'Planning for the 1980s: unemployment', *Planning 379*, 6–7.

Townsend, A.R. (1981) 'Geographical perspectives on major job losses in the UK, 1977–80', *Area 13* (1), 31–8.

Townsend, A.R. (1983a) *The Impact of Recession*, London, Croom Helm.

Townsend, A.R. (1983b) 'The use of redundancy data in industrial geography', in Healey, M.J. (ed.) *The Data Base for Urban and Regional Industrial Research in Britain*, Norwich, Geo Books.

Townsend, A. and Peck, F. (1985) 'The geography of mass redundancy in named corporations', in Pacione, M. (ed.) *Progress in Industrial Geography*, London, Croom Helm.

TUC (1982) *Regional Development Planning, A TUC Policy Statement*, London, TUC.

Urry, J. (1981) 'Location, regions and social class', *International Journal of Urban and Regional Research 5*, 455–74.

Wainwright, H. and Elliott, D. (1982) *The Lucas Plan: A New Trade Unionism in the Making?*, London, Allison & Busby.

Ward, T. and Rowthorn, B. (1979) 'How to run a company and run down an economy: the effect of closing steel-making in Corby', *Cambridge Journal of Political Economy 3* (4), 327–40.

Watts, H.D. (1980) *The Large Industrial Enterprise*, London, Croom Helm.

Watts, H.D. (1981) *The Branch Plant Economy*, London, Longman.

Wear Valley District Council and Durham County Council (1983) *Wolsingham Steel Works: The Case Against Closure.*

Yemin, E. (ed.) (1982) *Workforce Reductions in Undertakings*, Geneva, International Labour Office.

Index

We have structured the index to help guide you through some of the arguments. Entries concerning research design and methodology are in **bold** type. Some very general words have been excluded – there would have been too many references – 'explanation' is a case in point. In such cases, follow up sub-headings such as 'causation', 'conceptualization', etc. We have put in a lot of links between entries, but others you should be able to follow up for yourself.